666 SCIENCE TRICKS & EXPERIMENTS
BY BOB BROWN

666 SCIENCE TRICKS & EXPERIMENTS
BY BOB BROWN

TAB TAB BOOKS Inc.
BLUE RIDGE SUMMIT, PA. 17214

FIRST EDITION

SECOND PRINTING

Copyright © 1978 by TAB BOOKS Inc.

Printed in the United States of America

Reproduction or publication of the content in any manner, without express permission of the publisher, is prohibited. No liability is assumed with respect to the use of the information herein.

Library of Congress Cataloging in Publication Data

Brown, Bob, 1907-
   66 science tricks & experiments.

    Includes index.
   SUMMARY: Instructions for performing experiments using readily available materials that demonstrate various scientific principles.
    1. Science—Experiments. 2. Scientific recreations. [1. Science—Experiments. 2. Experiments. 3. Scientific recreations] I. Title.
Q164.B8412   502'.8   77-19345
ISBN 0-8306-7881-6
ISBN 0-8306-6881-0 pbk.

DEDICATION

To W.M. Woods, first-rate scientist, perfect gentleman, astute writer, warm personal friend, and my most merciless critic. His observations—and arguments—have been of inestimable value in the preparation of this book.

Preface

This is a collection of the "Science for You" experiments that have run in newspapers through the L.A. Times Syndicate. There have been other collections, but this is by far the largest and most comprehensive ever assembled.

The book brings interest and delight to young and old alike who have a natural and investigative curiosity about how and why things happen.

This giant volume can be especially rewarding to teachers and students of science through junior high, fathers and mothers whose kids always need something challenging to do, prospective Science Fair contestants searching for good project ideas, and, not the least important, young people who seek desperately to understand the world around them. Many, if given the chance and preparation, will be the scientists and engineers of tomorrow—who just may save mankind from itself.

Many of the experiments are thought to have originated with the author. Many others are improvements and simplifications of existing experiments. Some are corrections of errors found in the explanations of experiments in other sources. The author has sought to minimize the possibility of error; he has three erudite and meticulous consultants who check everything.

Observant readers will notice that there is repetition in many of the explanations. The author, rather than rewrite to eliminate this, chose to present each experiment with its own full explanation, as it appeared in the newspaper column. Thus each experiment is complete in itself. Such repetition many be especially noticeable in Chapter 9, where there are many Bernoulli experiments.

Comments and suggestions are always welcome as the author presents more and more experiments through his column.

Bob Brown

Contents

Foreword 6

1 Inertia & Momentum 7

2 Sound & Other Vibrations 15

3 Projects to Build 38

4 Tricks 69

5 Biology & Psychology 99

6 Water & Surface Tension 150

7 Gravity & Centrifugal Force 197

8 Electricity & Magnetism 202

9 Air, Air Pressure & Gases 238

10 Heat 273

11 Light 291

12 Household Hints 314

13 Chemistry 331

14 Mechanics 371

Index 402

Foreword

This book fills an important need in our science-conscious, space-minded era. Not only does it lead many youngsters into the enchanting world of science, it also introduces elementary and junior high pupils to many original experiments which are not yet in standard texts.

Mr. Brown's experiments run first in his valuable newspaper column, "Science for You" (syndicated by the Los Angeles Times Syndicate). Both the column and the other books he has written find their way into classrooms, where teachers and students alike have been delighted with the simplicity of the explanations and the ease with which they can obtain the necessary substances and gadgets.

The books and column have become popular in many homes where adults as well as children have found many hours of pleasant reading and experimenting in them. The experiments appeal to our curiosity and our eagerness to learn—traits which have no age limits.

David Dietz
Science Editor Emeritus
Scripps-Howard Newspapers

Chapter 1

Inertia & Momentum

THE ELASTIC COINS

NEEDED: Five nickels and a copper penny.

EXPERIMENT 1: Flip one coin against another, as in the upper drawing, and the elasticity of the metal will be demonstrated by the bouncing of the coins. Flip the penny hard against the pile of nickels, and one nickel will fly out from the bottom of the stack as shown below.

EXPERIMENT 2: Try a quarter. If the quarter stikes head-on and at the right velocity, the bottom nickel flies out and the rest of the nickels come to rest on the top of the quarter. If the quarter strikes at a grazing angle the nickel flies out at the bottom and the quarter goes off at an angle.

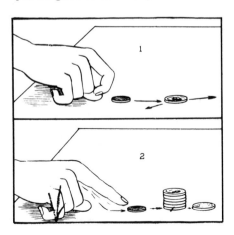

The Elastic Coins

REASON: The coins are not pure soft metal, but alloys that increase hardness and the elasticity. Therefore, they bounce. The size of the coins is important. The "copper penny" referred to here is not an English penny, but the standard American cent piece which is part copper. The penny must be thinner than a nickel, so as not to hit the second nickel in the stack. A dime will work if it is flipped hard enough.

In the second experiment, the inertia of the stack of nickels keeps it practically still while the bottom coin is flipped out from under the pile.

Simple experiments such as these can involve mathematics if the experimenter wants to go that far into them. Other words that may be used in their interpretation can include impact, impulse, and conservation of momentum.

A DOLLAR BILL PUZZLE

NEEDED: A dollar bill, an empty bottle, and a pencil.

EXPERIMENT 1: This is one of the old-time puzzlers: take the bill from under the bottle without touching the bottle or tipping it over. It's easy if the bill is rolled around the pencil and the bottle pushed aside in that way.

EXPERIMENT 2: Try jerking the dollar bill quickly while the bottle rests on it. The bottle will probably not tip over; its inertia tends to hold it in place. Do not raise the bill as you jerk.

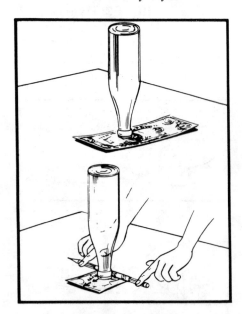

A Dollar Bill Puzzle

EXPERIMENT 3: Pull the dollar bill slowly. The bottle will probably move with it. Pull a little more quickly, and the bottle will probably tip over.

REASON: A light push on the upper part of the bottle will create enough torque to turn the bottle over. It is not very stable in this inverted position because of its high center of gravity.

A similar push at the lower part of the bottle causes a much smaller torque but causes enough force to slide the bill from under the bottle.

Soft drink bottles are good for these experiments. They must be dry.

CENTER OF GRAVITY

NEEDED: A long pencil and a short one.

EXPERIMENT 1: Stand the pencils upright, and release both at the same time. The short one falls to the table much faster than the long one.

EXPERIMENT 2: Notice that when the pencil has fallen and is horizontal it moves in the direction of the fall. The mass center has some horizontal velocity because the pencil was forced to fall on the arc of a circle.

EXPERIMENT 3: Watch a small child fall from a standing position. He almost never gets hurt. The fall is so short that the speed gained is not great; he does not hit the floor very hard. If he was much taller, or fell from a table, gravity would have more time to accelerate him (speed him up), and he might hit the floor hard enough to hurt himself.

REASON: Gravity is the force which acts on both pencils to bring them to a horizontal position. The center of gravity of the longer pencil is much higher above the table than that of the shorter; therefore, in the longer pencil, more mass must fall through a longer distance with a longer arc. This takes more time.

BALANCE A HATCHET

NEEDED: A hammer, hatchet, axe or garden hoe.

EXPERIMENT 1: Try to balance the hatchet with the handle up. It is difficult if not impossible. Try balancing it with the heavy end up. It is easy.

EXPERIMENT 2: Try this with a broom or a long pole. It is much easier to balance a long object for the same reason. Circus performers take advantage of this principle when they balance one another on chairs on long poles.

EXPERIMENT 3: Note that the circus wire walker balances himself with a long pole that bows down at the ends. Try this (near the ground!) and see if it is easy. If the pole bows down far enough it will be self-balancing.

REASON: The hatchet will try to fall in either case, but if the heavy end is up, there is more interta to hold it in position, so you have more time to move the hand to counteract the falling movement.

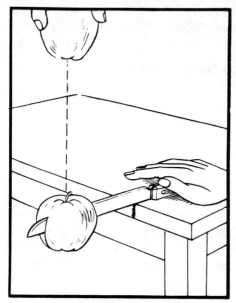

Mass, Velocity, Momentum

MASS, VELOCITY, MOMENTUM

NEEDED: An apple, a sharp butcher knife, a table top.

EXPERIMENT 1: Hold the knife securely on the table, and drop the apple on it from a height of several feet. The apple will cut itself as it hits the blade.

REASON: The momentum of the apple as it falls will carry it past the blade, even though the friction offered by the blade as it cuts the apple will slow it down considerably.

Mass times velocity equals momentum. The farther the apple falls, the greater its momentum becomes.

A BRICK TRICK

NEEDED: A brick and a hammer.

EXPERIMENT 1: Hold the brick above the ground. Hit it with the hammer, and it will break without hurting the hand.

EXPERIMENT 2: Try breaking a half of a brick this way. It is much more difficult. The smaller mass of the brick presents less inertia. More of the force of the hammer blow goes through the small object to the hand. The half brick, or "bat," is about 16 times as difficult to break in the hand.

REASON: The muscles of the hand, arm, and shoulder are flexible and descend slightly as the hammer strikes the brick. The inertia of the brick would tend to keep it at rest, but the momentum of the moving hammer would tend to give the brick a high velocity. The result is that the brittle trick breaks by moving faster where the hammer strikes it than anywhere else.

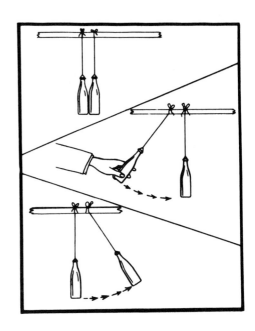

Bumping Bottles

BUMPING BOTTLES

NEEDED: Smooth-sided bottles suspended on strings so they barely touch each other.

EXPERIMENT: Let one bottle swing against the next. The first bottle will stop, and the motion will be transferred to the last bottle in line. This can be done with two or more bottles.

REASON: Laws of conservation of momentum and elasticity are demonstrated here. The bottles are elastic, allowing the momentum to be transferred through them.

Use thick bottles. They are not likely to break, but be careful of glass.

This works better if the bottles can be suspended by two bridles instead of single strings. Two bridles will confine the motion of the bottles so they do not swing in unwanted directions.

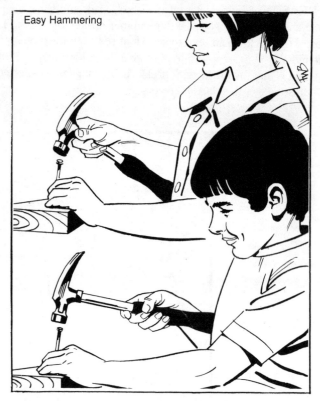

EASY HAMMERING

NEEDED: Hammer, nails, wood.

EXPERIMENT: Hold the hammer close to its head and try to drive a nail with it. Then hold it by the handle, farther from the head, and see how much easier it is to drive the nail.

REASON: When the hammer is held by the handle, farther from the heavy head, the velocity of the head is greater when it strikes the nail. This is associated with greater kinetic energy of the head, making it possible to transfer more energy (work) to the nail with each blow, while driving it into the wood.

"Inertia" is a term applicable here. The hammer, in motion, tends to continue in that motion until stopped by the impact with the nail.

The Loose Handle

THE LOOSE HANDLE

NEEDED: A hammer or other implement with a loose handle.

EXPERIMENT: Try to put the handle into the implement tightly by hitting the heavy head itself. Then try hitting the handle. The hammer tightens when the handle is hit in line with its long axis.

REASON: The heavy iron has inertia—more inertia than the lighter wooden handle. As the handle is hit the energy of the blow is transmitted through it to the hammer, but the hammer's inertia prevents it from moving as quickly as the energy in the handle reaches it. So the handle is driven in.

Inertia is that property of matter which manifests itself as a resistance to any change in the motion of a body (McGraw-Hill Encyclopedia). The inertia of a body is proportional to the mass of the body.

A simplified definition is that if the body is still it tends to remain still; if it is in motion it tends to remain in that motion. This is Newton's first law of motion.

MUSCULAR MOLECULES

NEEDED: String, a weight such as thread spool, newspaper, sticky tape or glue.

EXPERIMENT 1: Attach the spool to the string. Cut paper into a strip, and glue it to the spool or upper part of the string. Whirl

Muscular Molecules

the weight (out of doors on a calm day) and the spool will pull outward as far as the string will let it reach, while the paper will form an arc, extending itself very little farther than the spool from the hand.

EXPERIMENT 2: Note how easy it is to turn the page of a newspaper by moving the hands quickly. Inertia of the air beside the paper holds it as the turn is made.

REASON: As the paper moves around with the spool and forms an arc, it is bombarded by air molecules from all sides. The pressure of these molecules keeps the paper in the path opened by the spool.

We live in a great and heavy mass of air, but hardly notice it unless the wind blows hard or we move fast.

And where does inertia come in? Inertia is the tendency of a body to stay at rest or continue at the same speed and in a straight line unless an outside force is applied. Here, there is a slight force tending to push the paper strip outward, but little or no other force to move the strip out of its circular path as it slices its way through the air.

Chapter 2

Sound & Other Vibrations

HOW FAR THE STORM?

NEEDED: A distant thunderstorm (or other source of sound) and a stop watch.

EXPERIMENT: Start the count as the lightning is seen, and stop when the sound of the thunder is heard. The length of time taken for the sound to reach the ear tells the distance of the lightning flash.

REASON: Sound travels in air about 1100 feet per second. If it takes five seconds for a sound to reach the ear, the distance is about 5500 feet, a little more than a mile.

The light from the flash travels so fast (more than 186,000 miles per second) that its speed need not be considered in the calculations.

A good way to estimate the distance if there is no stop watch is to count "thousand one, thousand two, thousand three," etc., so that each count would be a second, representing about one-fifth of a mile.

A TRICKY HUMMER

NEEDED: A cardboard tube and a tall jar of water.

EXPERIMENT: Place the tube in the water, and hold it with one hand while holding the jar with the other. Whistle or hum over the open end of the tube, keeping a constant note. Raise and lower the jar of water so that the water in the tube goes up and down slowly, thus making the air column in the tube longer and shorter.

When the correct point is reached, the sound is much louder.

A Tricky Hummer

REASON: When the point of resonance of the air column is reached, the air vibrates at the frequency of the whistle. At this point, each vibration of the air column reinforces the vibration of the whistle sound, making it louder. This is a state of resonance, and the vibrations are called "sympathetic" vibrations.

If resonance is not obtained from the first whistle or hum tone, try a different pitch. Also try a different position of the mouth at the end of the tube.

KNIFE AND FORK CHIMES

NEEDED: A knife, a fork, a spoon, some string.

EXPERIMENT 1: Tie the silver pieces together so that they do not touch each other. Hold the ends of the string to the ears, and

Knife and Fork Chimes

as the head is moved and the silver pieces clang together, there is the sound of beautiful chimes.

REASON: The sounds heard are very much like the sounds of ordinary clanking of the silver, except that each sound lasts longer since the silver is free to vibrate. The string conducts the sounds to the ears, making them louder and more mellow.

The mellowness is accounted for by the softness of the string, which filters out some of the harshness. The vibrations in the individual pieces are at regular frequencies, and these give musical tones. Irregular vibrations would make noise.

EXPERIMENT 2: Replace the string with rubber bands tied together. Hold the rubber to the ears as the string was held. There will likely be no sound at all, certainly not chimes. This is because rubber is not elastic in the scientific sense.

REASON: Vibrations from the silverware are fed into the string, and travel up the string to the ears. The string is elastic enough to transmit the vibrations with a little loss. In the rubber, the sound energy introduced in the stretching in each vibration is not completely given back in the relaxation, but some of it is turned into heat.

The sound wave gets weaker as it goes up the rubber, and soon dies out completely.

The common definition of "elastic" is "stretchable-but-finally-coming-back." So, in common usage, rubber and things woven of rubber are elastic. In the scientific sense, glass and hard steel are very elastic, while rubber is not.

Irregular Vibrations

IRREGULAR VIBRATIONS

NEEDED: A drinking glass, a pencil, a string.

EXPERIMENT: Tie the string tightly around the glass and loosely around the pencil. Hold the pencil and let the glass hang down. As the pencil is turned, a peculiar noise comes from the glass.

REASON: The string does not slide easily on the pencil when the pencil is turned, but turns with the pencil for a short distance, then slips back. The jerky movement makes vibrations in the string, and the string transfers them to the glass. The glass sets up vibrations in the air within and around the glass, and these reach the ear.

The Open and Shut Pipe

THE OPEN AND SHUT PIPE

NEEDED: A large-diameter soda straw and a sharp razor blade.

EXPERIMENT: Cut the straw at about the middle. Flatten one end of one piece of the straw. Arrange the pieces as shown and blow through the straw. A musical tone will be heard. Remove your finger from the end of the lower straw and blow again. The new musical tone will be an octave higher.

Try different lengths of the lower straw.

REASON: This is a condition of resonance (look up the word) of an air column in a pipe as in organ pipes or other wind instrument.

The air column in the closed tube vibrates a maximum amount at the open end (antinode) and none at the closed end (node). In the open tube there is maximum vibration at each end (antinode) with a node between them. The wave length in the closed tube is two times that in the open tube. Therefore, the frequency in the closed tube is one half that in the open tube.

A Musical Rubber Band

A MUSICAL RUBBER BAND

NEEDED: A tin can, two nails, a rubber band.

EXPERIMENT: Drive the nails halfway into the can, as shown, and stretch the rubber band between them. Pluck the band with the finger and a sound will be heard. Wrap the band around the nails a few times, pluck it, and the pitch heard when the band is plucked will be higher.

REASON: As the band is plucked, it is set into vibration. The greater the tension on the band, the faster the vibration, and the more rapid the vibration, the higher the pitch.

The can serves to hold the nails and to vibrate with the nails. The air in the can vibrates, too, and may make the sound louder.

HEAR THROUGH THE TEETH

NEEDED: A rubber band.

EXPERIMENT: Hold the band stretched between the teeth and a finger, pluck it, and a musical sound is heard. Have someone

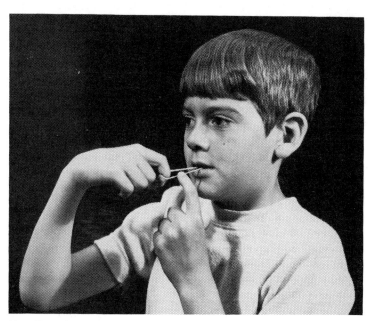
Hear Through the Teeth

else hold the end that was between the teeth, pluck it again, and the sound is not as loud.

REASON: The teeth and jaws act somewhat as a sounding board, vibrating with the vibrations in the rubber band. The sound waves are conducted directly to the ears through the bones of the skull. When the band does not touch the teeth only the sound waves produced in the air by the vibrating band reach the ears through the air, as in normal hearing. Also, the soft flesh of the fingers reduces the sound when hands are used instead of teeth.

A MYSTERY SOUND

NEEDED: A glass, a fork, a table top without a cloth.

EXPERIMENT: Have someone sit at the edge of the table with an ear to the drinking glass. Strike the fork. Touch the handle of the fork to the table top. Suddenly the sound of the vibrating fork seems to be coming form the glass.

REASON: If the fork is held above the table, a slight sound of the vibrating fork may be heard directly through the air. But if the fork is touched to the table top, the table vibrates much as the sounding board of a musical instrument. The vibrations are carried from the fork to the table, to the glass, and into the air surrounding the glass, where the sound seems much louder.

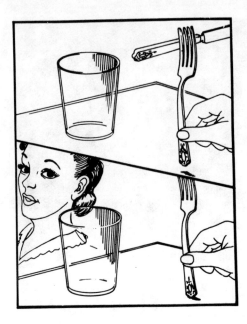

A Mystery in Sound

Holding the ear over the glass makes the sound louder than if the ear is held an equal distance above the table because of resonance vibrations of the air column in the glass itself. Different-sized glasses may be tried.

WHAT CAN CARRY SOUND?

NEEDED: A table top, a bucket of water, a comb.

EXPERIMENT 1: Put your ear on the table, and have someone tap the table at the other end. The sound is carried to the ear through the table top.

EXPERIMENT 2: Put your ear to the bucket of water, and have someone rub his fingers over the teeth of the comb while it is under the water. The sound will be carried through the water to your ear.

REASON: Most sounds come to us as vibrations in the air. But they can come through many substances—not only water and wood, but through the bones of the head and our teeth as well. They travel least easily through soft, sound-deadening substances.

MUSICAL BALLOON

NEEDED: A rubber balloon.

EXPERIMENT: Blow up the balloon. Place it under the arm and pull the neck of the balloon as shown in drawing at upper left.

By varying the pressure and the stretch, musical (?) tones can be produced.

REASON: Air escaping through the stretched rubber neck of the balloon will not flow steadily, because the rubber expands and contracts, causing the air to come out in a series of puffs or waves. If these are irregular, they produce noise. If they come in regular intervals, they can make rather pleasant muscial sounds.

Musical Balloon

SODA-STRAW MUSIC

NEEDED: Large-diameter straws, scissors.

EXPERIMENT: Press the straw end with the fingers. With a little practice, you can blow into the straw and produce a musical note. Clipping off the straw makes the pitch higher.

REASON: Practice is necessary, as the lips must be taught to vibrate properly. This is necessary in the playing of some of the musical wind instruments.

The length of the straw controls the length of the air column in the straw. The pitch is varied by the length of the vibrating air column.

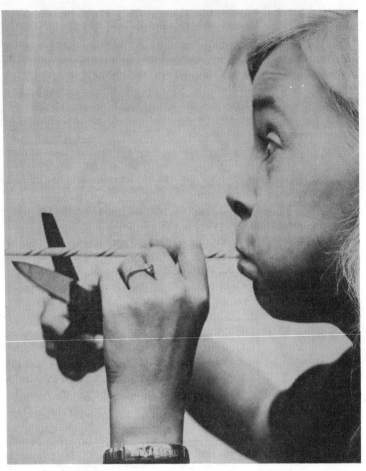

Soda Straw Music

THE NOISY CAN

NEEDED: A tin can, a string, rosin, a pencil, a matchstick.

EXPERIMENT: Make a hole in the middle of the can, on the bottom, and tie the string on the matchstick inside the can. Have someone hold the can while you draw the rosin back and forth on the string.

The rosin will make an unpleasant sound. When the pencil is then drawn along the string, it, too, will make the unpleasant sound. Remember that a similar movement of the violin bow across the strings can make beautiful music.

REASON: The rosin sticks, releases, sticks, and releases time after time. This makes a jerky movement in the string, which is transmitted to the can as irregular vibrations. The rosin makes the pencil move at irregular intervals over the string.

The vibrations of string, pencil, and can are transmitted through the air to the ear, and are noise. In the tuned violin, the vibrations come at regular intervals, and so can be music!

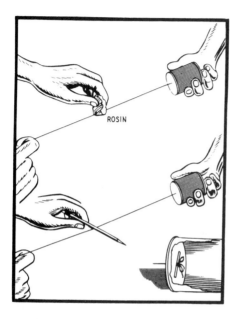

The Noisy Can

THE BULLDOG

NEEDED: A piece of thin wood such as a ruler, a strong cord, a round stick.

EXPERIMENT: Tie the cord to a hole in the end of the ruler. Tie the other end of the cord to a notch in the round stick—loosely so it can turn on the stick. Whirl the ruler around overhead, and listen to the bulldog growl!

REASON: As the thin wood moves through the air it is caught by air currents and twirled. As it twirls it creates disturbances in the air that reach the ear as regular but rather discordant sounds.

The tone is changed with the speed at which the wood is twirled, and is different with different shaped pieces of wood. The operation of this little toy is complex, and its explanation could start many an argument in a college physics class.

Be careful the wood does not fly out of control and hit someone.

Mr. Doppler's Effect

MR. DOPPLER'S EFFECT

NEEDED: While driving on the highway, notice that an automobile horn has a higher pitch as it approaches than when it has passed and is speeding away.

REASON: Sound is made up of waves in the air. When we are approaching the horn, our ears receive more of the waves per second than the horn is producing; if we are receding, our ears will receive fewer per second. This is called the "Doppler principle" because of its discovery by Christian Doppler more than a hundred years ago.

AN AIR HORN

NEEDED: A soft drink bottle in a moving automobile.

EXPERIMENT: Hold the bottle at the car window, changing its position. At one point a whistle will be heard coming from the bottle.

REASON: Air flowing across the opening of the bottle makes a whistle just as a sound may be heard by blowing the breath across a bottle neck opening. Air blown into the bottle mouth swirls into a vortex, the action alternately compressing and expanding the air. The compressions and expansions set up waves of air outside the bottle, and these comprise the sound that reaches the ear.

An Air Horn

TUNING THE DRUM

NEEDED: Observation of a large orchestra.

EXPERIMENT: Observe a large orchesta. The story has it that a small boy, watching the tympani player in the orchestra, wanted to know why the tympanist smells the drum occasionally.

REASON: The musician does not bend close to the drum to smell it, but to listen to it as it is lightly struck with a finger. The head, stretched over the drum, is a membrane that is affected by humidity of the air, and it gets out of tune very easily. The player checks it several times usually in one concert. By holding the head close it is possible to hear the minute sound of the drum and tune it while the other instruments of the orchestra continue to play.

VIBRATIONS THROUGH A CONE

NEEDED: A phonograph and a 78 RPM record, a sheet of stationery, a pin, cellophane tape.

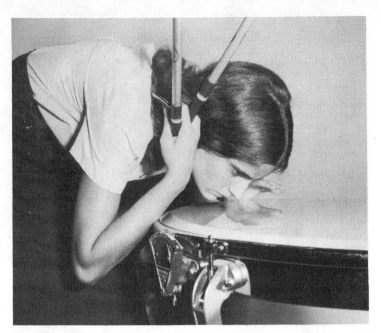
Tuning the Drum

Vibrations Through a Cone

EXPERIMENT: Make a cone of the paper, stick the pin into it as shown, hold the pin in the record grooves as the turntable turns, and the music will be heard coming from the cone.

REASON: A phonograph groove is not straight, but is full of waves. As the waves pass against the needle they set the needle to waving, or vibrating, and the vibrations are transferred to the paper. Since the paper is larger than the needle or pin it sets the air to vibrating more than would the pin alone, and the vibrations are heard by the ear.

A 78 RPM record is necessary for this for two reasons. First, the grooves and waves in a 45 RPM or long-play record are too small to register very loudly, and second, the grooves are so small and fragile that an ordinary pin or needle can damage them.

HEARTBEATS THROUGH A LOUDSPEAKER

NEEDED: A strong public address amplifier, a two-inch loudspeaker, shielded cable, plug.

EXPERIMENT: Solder the plug on one end of the cable, making sure the shield is attached to the correct place. Solder the loudspeaker to the other end. Plug it in, turn up the volume, touch the loudspeaker to the bare chest over the heart, and the beating can be heard.

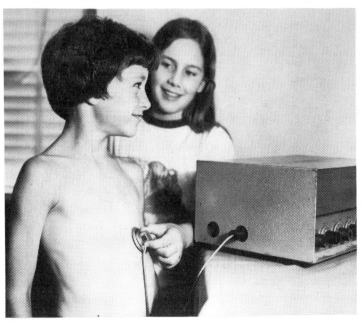

Heartbeats Through a Loudspeaker

REASON: The loudspeaker consists mainly of a movable diaphragm attached to a coil of wire that can move in the lies of force of a permanent magnet. Each movement induces a small current in the coil, and this current, transmitted to the amplifier, is changed into vibrations in the air which are heard.

It is necessary to put the public address loudspeaker some distance away to prevent feedback. Ten feet should be sufficiently far away.

Almost any school will have a strong amplifier. More volume may be produced by adding a pre-amplifier. This can be built from an electronics store module costing about $2.

VOICE VIBRATIONS

NEEDED: A can, a rubber sheet (may be cut from a balloon), a piece of mirror, a light colored wall or sheet, sunlight.

EXPERIMENT: Tie the rubber over one end of the can (both ends of the can must be cut out), and glue the small piece of mirror on

Voice Vibrations

the rubber, not in the middle. Place the can on a post or hold it very still, adjusting it so sunlight will reflect a small spot from the mirror to the wall. Talk into the open end of the can.

REASON: The vibrations of the voice transmitted through the air in the can cause the rubber to vibrate. The small motions of the vibrating rubber are amplified by the light reflection from the mirror. The pattern on the wall will suggest the pattern shown on the screen of an oscilloscope, an electronic device that shows pictures of electrical and other phenomena on a screen similar to a television screen.

A large coffee can works well for a deep adult voice. A slim frozen juice can works better for a child's or woman's voice.

MAKE YOUR OWN DRUM

NEEDED: Coffee can, rubber balloons, a funnel, a lighted candle.

EXPERIMENT: Both ends of the can must be cut out. Stretch balloon rubber over both ends, and hold the rubber with bands or strings. Tap on one end and the rubber sheets will vibrate—it is a small drum!

Make Your Own Drum

REASON: The vibrations of the sheet struck with the hand continue through the air in the can, making the other sheet vibrate. If a funnel is held as shown it can concentrate the moving air vibrations so the air from the funnel can snuff out a small candle flame.

The wave travels very fast through the can. A wave traveling through pipe and hose in a long railroad train can reach all the cars in a very short time, setting all the brakes.

SYMPATHETIC VIBRATIONS

NEEDED: A piano.

EXPERIMENT 1: Uncover the strings of the piano by opening the lid. Hold the loud pedal down, releasing the strings so they can vibrate freely. Sing a note, loud, into the piano, and the sound can be heard coming back for a few seconds as a string tuned to the pitch of your voice continues to vibrate.

EXPERIMENT 2: Try other musical instruments to make the piano strings vibrate.

EXPERIMENT 3: If two pianos can be found in the same room, see if one of them can set up sympathetic vibrations in the other.

REASON: Sound is made up of vibrations (in air, in this case) and the vibrations falling on a piano string tuned to the pitch of the voice sound will begin to vibrate "in sympathy." The string will continue to vibrate after the sound that started it cannot be heard.

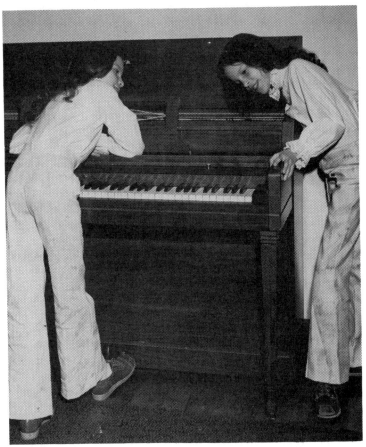

Sympathetic Vibrations

A MEGAPHONE

NEEDED: Light cardboard.

EXPERIMENT: Roll the cardboard into megaphone shape. Talk through it to someone several feet away, then talk in the same voice without it. The voice carries farther and is louder when the megaphone is used.

REASON: In ordinary speech the sound waves go out in all directions, getting weaker as they go farther. The megaphone concentrates them so that more of them go in one general direction without losing so much of their energy in that direction.

A Megaphone

THE GHOST WHISTLE

NEEDED: Two whistles such as those used on tea kettles. The whistles must give steady notes. A police type whistle which has a ball inside, giving it an intermittent sound, cannot be used.

REASON: The whistles should have almost the same pitch but not quite. Have two people blow the whistles and a "beat" will be heard. The beat is heard as pulsations which represent the difference in frequency of the two sounds made by the whistles. Have the people blowing the whistles stand at different distances from each other and the effect can be changed somewhat. Try this by having two people whistle with their lips. Many effects can be produced.

The Ghost Whistle

BIG SEA, LITTLE SEA

NEEDED: Two jars of different sizes.

EXPERIMENT: Everyone has heard that if a seashell is held to the ear the sounds of the sea may be heard. Here we see that jars serve the same purpose, but a deep heavy "sea" is heard in the big jar, and a light, higher-pitched "sea" sound is heard in the smaller jar.

REASON: The sounds heard are "sympathetic" vibrations, which means that they are sound waves in the room or out of doors that in turn set up waves inside the jar. The air in the jar can vibrate normally at certain pitches, depending on the size and shape of the jar. A little air in a little jar can vibrate at higher pitches than more air in a larger jar.

PENDULUMS

NEEDED: A spring from an old window shade roller, some nails or other weight, some string.

EXPERIMENT: Suspend the spring and the string. Attach weights to each. The weight on the string will swing back and forth in

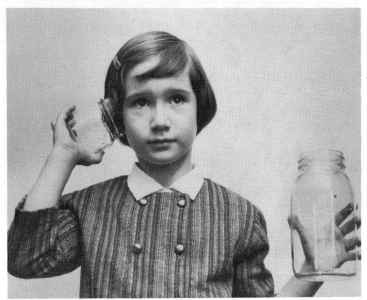

Big Sea, Little Sea

the manner of the pendulum. Its path is not always in one dimension, but the outline of the swing is a long narrow ellipse. The weight on the spring may be made to go up and down or to twist. The outline of its swing may also be an ellipse. This gives us three types of pendulums.

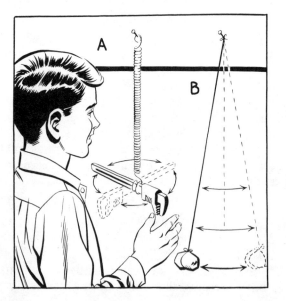

Pendulums

Waves Shown by "Slinky"

WAVES SHOWN BY "SLINKY"

NEEDED: A "Slinky" spring and a flat slick table top.

EXPERIMENT: Hold the spring as shown, flick it with the finger, and a wave will travel through its length. The wave may be seen very distinctly.

REASON: As one coil of the spring is moved, it transfers its energy of movement to the next coil then returns to its former position. The second coil does the same, then the third, etc., until the wave has traveled the length of the spring.

The same thing happens in the air when a sound is produced, except that the wave travels from the source in all directions radially. A small amount of air is moved; it transfers its energy to the air next to it, and so on. Air with sound waves moving outward is in a state of alternating invisible condensations and rarefactions of its molecules.

Chapter 3
Projects to Build

A VIBRATION DETECTOR

NEEDED: A good flashlight cell, three dead cells, pencil sharpener, earphones, wires and connecting clips, wood, woodworking tools.

EXPERIMENT: Take a carbon out of a dead cell, and sharpen it at both ends. Bore holes in the other two carbons from the dead cells, so they will hold the sharpened one when set into dowels. The sharpened carbon must rest loosely in the holes. Connect as shown. One flashlight cell will give enough power.

OBSERVATION: Set the instrument on the ground, and ground vibrations may be heard. If the earphones are in another room, speech may be heard coming over the device.

REASON: Air movement or vibration causes the carbon-to-carbon contact to be better or worse, varying the amount of current that may pass through it. This the principle of the telephone microphone, which is called the "transmitter." Carbon granules are used in it. (Idea suggested by Russell Richner, Lake Worth, Florida.)

THE INDIAN DRILL

NEEDED: A broom handle, a nail, a string, pieces of wood, some tools.

EXPERIMENT: Assemble the pieces as shown in drawing at left so that the wooden handle will be stopped by the string just

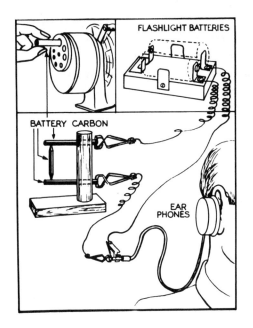

A Vibration Detector

above the flywheel. Start the drill by winding some of the string as shown in drawing at the right. By moving the handle up and down, the drill will be made to spring rapidly back and forth. The nail will drill a hole in the wooden block.

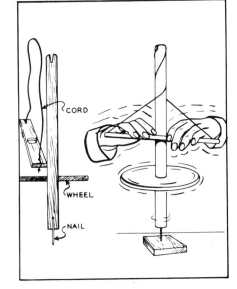

The Indian Drill

REASON: The principle of this instrument is similar to that of the yo-yo, in that the pulling of the string starts the instrument in a rapid turning motion. Then its momentum will carry it on, while the string winds around the broomstick in the other direction.

This device has been used by many primitive tribes in many parts of the world to drill holes and start fires. Because it was used by some of the American Indians it has been called "The Indian Drill."

A MAGIC PROPELLER

NEEDED: Wood, a screw, a knife.

EXPERIMENT: Carve a stick so that the notches on it are evenly spaced. They do not have to go all the way around the stick; the trick will work if they are cut on only two edges. Mount the propeller on the end with the screw, so that it is loose and turns freely.

Rub one stick with the other, while rubbing the finger against the notches. The propeller turns. Place the finger on the other side of the notched stick, and the propeller reverses and turns in the opposite direction.

REASON: If the notches are rubbed with the stick alone, the vibrations in the stick are likely to be straight back and forth, as shown in small diagram A. The propeller will not turn.

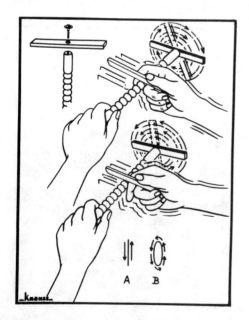

A Magic Propeller

When the stick and fingers both rub the notches, the vibration takes a circular path, as in diagram B. The propeller turns because of this circular vibration. Practice makes perfect!

MAKE YOUR OWN ALCOHOL LAMP

NEEDED: A jar with a metal cover, a large nail, some cloth for a wick.

EXPERIMENT: Punch a hole in the lid with the big nail as shown. Drive the nail into wood from the upside-down lid, so that the projecting metal in the lamp will point upward. Put the wick in, put alcohol in the jar, and you have a lamp that will give a hot flame.

The metal lid will conduct heat away from the flame, so that the flame does not go below. the lid into the alcohol. But any fire is dangerous, so be careful.

Make Your Own Alcohol Lamp

A HOMEMADE TELEPHONE

NEEDED: Two paper cups, string.

EXPERIMENT 1: Make holes in the bottoms of the cups, put string through the holes, and tie short sticks to the ends to hold the strings in place. Stretch the string—you have an improved "tin can telephone."

A Homemade Telephone

This is an old experiment when tin cans are used. It will be found that paper cups work better, because the paper vibrates more easily than the metal of a can.

Use hard twisted string; soft cotton string will transmit less of the vibrations from one cup to the other.

EXPERIMENT 2: If fine wire is available, try it in place of the cord. If the wire is fine it should conduct the sound. If it is heavy its inertia will kill some of the vibrating.

MEASURE AIR MOISTURE

NEEDED: A board, a peg, some paper-backed foil, glue.

EXPERIMENT: Mount the peg in the board. Cut a strip out of the foil, glue the end of the strip to the peg, then wind the strip of foil around the peg.

REASON: The coil of foil will unwind somewhat, then will wind and unwind slightly more as the moisture content of the air varies. The paper in the strip absorbs moisture and expands. The foil does not. The paper will force the foil to bend slightly as it expands and contracts.

This crude instrument cannot, of course, be depended on as a hygrometer or an air moisture measuring device. But it illustrates the principle of one type of such device.

Measure Air Moisture

MAKE YOUR OWN KALEIDOSCOPE

NEEDED: A 5 × 7 dime store mirror, a way to cut it, plastic wrap or tissue paper, rubber bands, various colored soda straws.

EXPERIMENT: Cut the mirror lengthwise into three pieces (they will be about 1 5/8 × 7) and cut tiny pieces of soda straw with scissors. Hold the three glass pieces together (as shown) with rubber bands, and put the paper or plastic over one end. Hold it on with a rubber band.

Hold the device up to the eye as shown, turn it slowly, and many beautiful patterns will be seen in the plastic pieces as they fall— never two patterns exactly the same.

REASON: The explanation is the multiple reflections of the colored bits in the three mirrors. Each mirror reflects them at least two times.

BUILD YOUR OWN SCALE

NEEDED: Wood, a can lid, a hacksaw blade, screws, string.

EXPERIMENT: Make the scale or balance as shown, a hacksaw blade as the spring (suggested by the Nuffield Junior Science Apparatus book).

Make Your Own Kaleidoscope

To calibrate the author's model he used weights of one-half, one, two, and four ounces. It is suggested that metric weights be used if available.

Build Your Own Scale

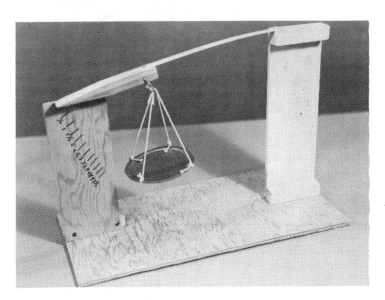

Build Your Own Scale

OBSERVATION: This scale is quite sensitive; its usefulness is determined mainly by the accuracy and exactness of the calibration.

An instrument such as this is often called a "balance." This is not technically correct. A balance compares masses hung on each end of a lever; no spring is used.

THE DIODE (1)

NEEDED: A diode, a lamp, socket, battery, all to match, a double-pole, double-throw switch, wire for connections.

EXPERIMENT: Connect the parts as shown in the diagram. When the switch is closed either way, the battery is connected to the lamp, but the lamp will burn one way and not the other.

REASON: A diode is a strange instrument that will let current flow in one direction through it, but not the other. Reversing the switch reverses the polarity of the battery, which means that it reverses the direction of the current flow.

Diodes, along with transistors, are the "solid state" devices that have revolutionized electronics in the past few years. They do tasks that formerly had to be performed with vacuum tubes.

Parts needed for this experiment may be bought from a radio supply store.

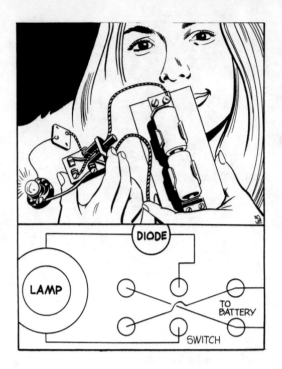

The Diode (1)

THE DIODE (2)

NEEDED: A diode, a lamp, a single pole, double-throw switch, a transformer, wire for connections.

EXPERIMENT: Connect the parts as shown, so that when the switch is closed in one position the current can flow directly to the lamp, but when closed in the other position the current must pass through the diode. The lamp is dim when the current has to flow through the diode.

REASON: A diode allows current to flow in one direction through it. Current coming from a transformer flows in one direction, then reverses and flows in the other direction, so about half the current is stopped by the diode and does not reach the lamp. The arrow points to the diode in the drawing.

Parts may be obtained from a radio and television store. Almost any diode will do for this, and they are not expensive. Some toy train transformers have rectifiers built in so that they give current that flows one direction (DC) and these will not work in this experiment. Straight (AC) current is required.

The Diode (2)

A FLYING SAUCER DETECTOR

NEEDED: A strong bar magnet, copper wire, brass screws, wood, tape, thread.

EXPERIMENT: Build the detector as shown. The magnet and some of the wire are put together by wrapping tape around them, as shown in No. 1. Part of the bare wire forms a loop or hook.

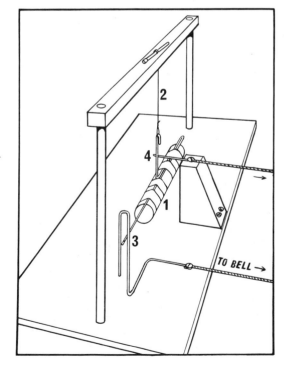

A Flying Saucer Detector

A thin thread, 2, hooks into the wire loop to suspend the magnet so it can turn freely.

A bare wire touches the hook or loop at 4, making electrical contact. At 3 is a loop that is touched as the magnet and wire turn slightly, making another electrical contact.

Set the instrument where it cannot be moved and where it will not get dusty (cover it!). There must not be a draft or breeze on it. Move it around until the magnet, pointing north and south, comes to rest so the extended wire is between the wires of loop 3, not touching. Keep all iron away.

OBSERVATION: A variation in the earth's magnetic field, caused by a flying saucer or a geomagnetic storm, will turn the magnet slightly, making contact at 3 and ringing the bell.

This instrument does not guarantee the appearance of a flying saucer; it does not mean that flying saucers exist. But it will make a good construction project.

(Adapted from a design by John Oswald, North Hampton. N.H.).

BELL ASSEMBLY FOR DETECTOR

NEEDED: Wires, bell, battery, switch, relay.

EXPERIMENT: Assemble the parts as shown. An electronics or electrical man will help, and this will be a good introduction to electrical circuit diagrams. The bell is shown at D, the relay at C, battery at B, and A shows the wires leading from the flying saucer detector previously mentioned.

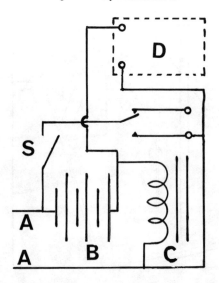

Bell Assembly for Detector

An electronics store is the place to get the parts, and the man there will help select them so they match.

OBSERVATION: When the flying saucer detector magnet is disturbed, it turns slightly, making contact to short out the wires shown at A and A. The bell starts the ring.

At the same time the relay operates to close its contacts, providing a permanent short to keep the bell ringing. The bell is silenced by opening the switch S.

As stated, this flying saucer detector is not guaranteed to detect a flying saucer. But it is an interesting project for a boy or girl for the science fair or just for fun at home or at school.

THE "PSYCHIC" MOTOR

NEEDED: Typing paper, a needle, a slender bottle.

EXPERIMENT: Construct the "motor" as shown, balancing it with the needle point resting on the top of the bottle. Hold the hand close to it, and it will begin to turn mysteriously.

The Psychic Motor

REASON: Martin Gardner, columnist for the *Scientific American,* presented this as a psychic motor. It was an April Fool joke, but fooled many people. He explains that the "motor" can be turned by air currents in the room, the breath, and convection currents caused by heat from the hand.

Gardner got the idea from a 1923 issue of *Science and Invention* magazine. The mystery motion is mysterious mainly because the motor is practically frictionless and can be run by imperceptible air currents.

POTATO POP-GUN

NEEDED: Potatoes, a metal or plastic tube with thin walls, a dowel, a knife.

EXPERIMENT: Cut potato slices a half to an inch thick. Press the tube down on them to cut out disks that will fit into the tube like corks. Place one in each end of the tube, push the dowel in *quickly,* and a potato "bullet" shoots out.

A Potato Pop-Gun

REASON: When the dowel is pushed in, the motion of the potato in the tube compresses the air between it and the potato disk at the other end of the tube. The compressed air forces the disk out, acting as a compressed spring.

The author, in building his pop-gun, used a piece of the thin overflow pipe found in toilet tanks. This came from a scrap box in a plumbing shop. For a dowel, he used the solid end of an old wooden window shade roller. The pipe had to be reamed to make the inside smooth at the ends where it was cut. The dowel should be nearly as large in diameter as the tube.

Many proprietors of plumbing shops are glad to help a boy or girl with a project in science.

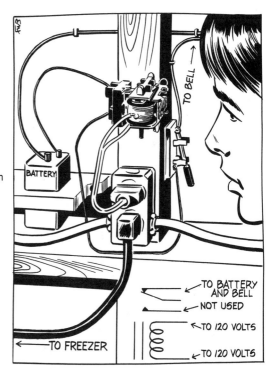

A Power-Off Alarm

A POWER-OFF ALARM

NEEDED: One relay, Porter & Brumfield GA11A, cord and plug, wire, bell, battery, switch.

EXPERIMENT: Connect the parts as shown. Keep the plug in. If the power goes off, the bell rings.

REASON: The electricity from the wall outlet keeps the relay energized. If the power fails the relay turns on the battery current to ring the bell.

If the GA11A relay is not found, a substitute will do, but get an electronics man to help with the connections. Be sure the relay is in a metal box, since its bare 120-volt terminals are dangerous.

Such an alarm might prevent tardiness at school or might prevent spoilage of food in a freezer if the power is off long enough. The bell may be turned off by opening the switch, but it will stop by itself when the current is on again.

A MEASURE OF DISTANCE

NEEDED: Wood bolts and nuts, a small spring, paint or crayon, felt or tape, glue.

EXPERIMENT: Build the device as shown in the drawing. Make the wheel 11.46 inches in diameter including a felt edge. The felt reduces slipping while in use.

The author used, as a handle, a 1 × 3 board sawed as shown. Plywood was used as the wheel. A painted mark or circle near the edge of the wheel shows a starting point for the wheel and allows visual counting of the revolutions. A click of the spring against a bolt with each revolution serves for auditory counting. Each revolution of the wheel signifies one yard distance.

Highway distances are frequently measured with such a device. (Suggested by the Nuffield Junior Science Apparatus book.)

A Measure of Distance

The Glass Harmonica

THE GLASS HARMONICA

NEEDED: An assortment of drinking glasses with stems, wood, screws, felt, water, vinegar.

EXPERIMENT: Hold a glass with the fingers of one hand securely on the table. Dip a finger into vinegar, rub it around the rim, and a musical tone should be produced. Try this with several glasses, adding or taking away water to adjust the pitch. The musical scale should be produced. Mount the glasses in line on a board. They may be mounted with three cleats to a glass. Felt should be glued on the bottom of the cleats to avoid cracking the flanges off the glasses. Two strips of wood are used to hold all the glasses in the row. Quarter-inch foam rubber is glued on the bottom of the strips to protect the glasses as the strips are tightened down with wood screws.

With nine glasses, all tuned, many beautiful melodies with harmonies may be played if two hands are used. Larger glass harmonicas, as used by professionals, have many more glasses, arranged in two rows. Four may be played at a time to produce four-part harmonies.

A BUZZER

NEEDED: Magnet wire, corner braces and screws from a hardware store, wood, tin can metal, a hack saw, soldering equipment, bolts and nuts, washers.

EXPERIMENT: Put washers on the bolts, put nuts on them, leaving room for them to go through the brace and be held with another nut. Wind magnet wire around them, making sure it is wound in opposite directions on the two bolts. Cut one brace off at

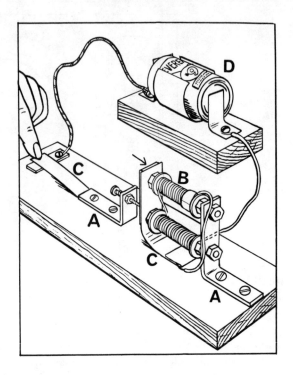

A Buzzer

the first hole. Put a thumbscrew through that hole. Attach both braces to the wood base as shown at A. First C is a switch made of tin can metal; next C is tin can metal screwed to the wood base so it can vibrate at its upper end. The arrow points to the piece of brace that was cut off—it is soldered to the can metal.

REASON: When the switch is closed, both magnets (bolts with wire) attract the can metal and the heavier iron that is soldered to it. When the attraction pulls the metal away from the thumbscrew in the first brace, the path of the current is broken, and the magnets allow the can metal and heavy iron to swing back, to make contact again.

The author wound 55 turns of No. 20 magnet wire around each bolt, as at B. The bolts used were quarter inch, two inches long. Braces were 1/8 inch thick, 2 1/2 inches to a side. This buzzer operates well on one D size flashlight cell.

Note that one wire from the magnets goes to the cell, the other to the tin can metal under the magnets. The magnets do not touch this metal.

A WAVE MACHINE

NEEDED: One piece of .048 steel wire 36 inches long, 54 pieces of 1/8 inch steel drill or welding rods, wood, screws, tin can metal, soldering or welding equipment.

EXPERIMENT: Make a form out of wood in which grooves are cut and equally spaced. The rods are placed in this for soldering. They are spaced 5/8 inch apart on centers. Make angles from tin can metal; cut v-grooves in them to support the steel wire on a plywood board. The steel wire must be free to turn in the grooves.

REASON: Move the end rod up and down, and the "wave" will travel to the other end of the assembly, reflect there, and return. Move the end rod up and down regularly, and a standing wave is produced. Attach an Erector motor to the end rod, so that an arm moves the rod up and down regularly, and the standing wave is made without the touch of hands.

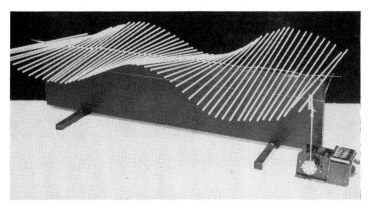

A Wave Machine

A MAGNETIZER

NEEDED: A cardboard tube from bathroom tissue, tape, rubber or plastic covered house wire, size 14, green twisted lamp cord (some lamp cord is too small), a male and female plug, switch, electric iron or heater, and soldering equipment.

The author stuffed dowels into the tube, and fitted plywood squares on the dowels to aid in the winding, as shown in photo A. Tape running through the tube with the dowels is used to hold the turns of wire together after the winding is complete. Push the squares against the ends of the tube, and start winding, using the No. 14 wire. Wind about 120 turns of the wire around the tube, tie

the tape around it to hold it, then remove the dowels and squares. Put more tape through the coil and tie it, as shown in photo B.

Connect the lamp cord as shown in photo C. A male plug goes on one end of it, the female on the other, and the switch in the center. Be sure to solder and tape the three joints in the wire. House current can kill. But this magnetizer is perfectly safe if the joints are well soldered and taped. To magnetize a steel object, insert it into the coil as shown in photo D. The electric heating device, connected as shown, draws enough current through the coil to make it a rather strong magnet, and its pull can be felt in the screwdriver or other object held in it.

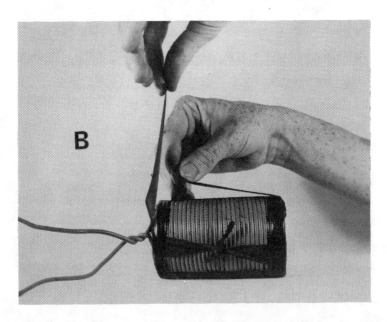

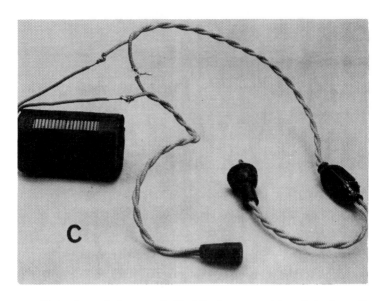

Turn the switch on *and off* while the screwdriver is in the coil, and it will become magnetized. Test it by picking up paper clips or nails. To demagnetize, put the screwdriver into the coil, turn on the switch, and remove the screwdriver *while the current is on.*

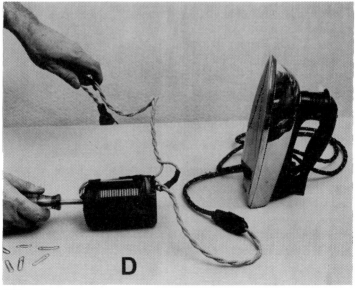

A Magnetizer

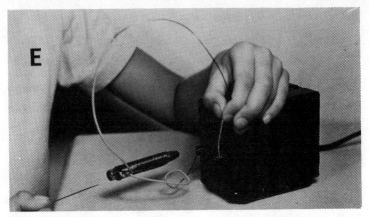

A Smaller Magnetizer

A SMALLER MAGNETIZER

A smaller magnetizer-demagnetizer may be operated from the safe current given by a toy train transformer. It can magnetize needles, and demagnetize them, as the larger one can handle larger steel objects.

A small non-metallic tube is needed. The author used the barrel of a cartridge type fountain pen, and wound 64 turns of magnet wire around it. Connect one end of the wire to the transformer, and hold the other end in the hand so it may be touched to the other transformer terminal. A switch is not needed.

Alternating current must be used.

REASON: House current is "alternating" current. That means it reverses its direction, and in the United States the reversals are 120 per second. Each time the current flows in one direction in the wires it magnetizes the tool in the coil in one direction. When the current reverses, the magnetism in the tool is killed, and takes place in the other direction. What was formerly a south pole becomes a north pole, and vice versa.

When the current is turned off while the tool is inside the coil the tool retains some of its magnetism of the moment. There is no way to know which end will be north and which end will be south.

When the tool is withdrawn slowly while the current is on the magnetism induced in the tool becomes weaker and weaker as it is reversed and the tool is moved farther and farther from the inside of the coil. By the time the tool is entirely out of the coil the magnetism remaining in it is so weak that it cannot be easily detected.

Common electric current in the United States is 60 cycle alternating. This means, according to the definition in *The Basic Dic-*

tionary of Science (Macmillan), that the current changes from its greatest value in one direction to its greatest value in the other direction and back again 60 times per second. The word "cycles" is being replaced by the word "hertz," which is abbreviated "Hz."

ON-SHORE AND OFF-SHORE WINDS

NEEDED: Cardboard box, with pan or dish of water to fill half of it, dark soil, charcoal, or wallboard painted black to fill the other side, incense for smoke, a photoflood or heat lamp, a knife, transparent plastic wrap.

EXPERIMENT: Leave a peep-hole in the side, covered with the plastic sheet. Cover the top with the plastic to keep out room air movements. Have the water slightly warmer than room temperature at the start of the experiment.

REASON: When the incense is lighted, and the box closed, the smoke will move over the water, because the warmer water warms the air above it, and starts the air moving up, then down toward the dark soil. Turn on the light. As the soil becomes warmer than the water, the air movement is reversed. The smoke will blow slowly toward the soil, up, over the other side of the box, then down over the water. This shows how air currents blow from the ocean to land when the land is warmer than the water.

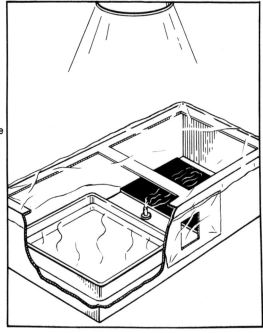

On-Shore Off-Shore Winds

Make a Light Engine

MAKE A LIGHT ENGINE

NEEDED: Plywood or stiff cardboard, compass for drawing circles, rubber bands, a sewing machine bobbin with holes (from dime store), nail, screws, wood, tools, a heat lamp.

EXPERIMENT: Assemble the engine as shown, so the plywood ring is supported on the bobbin with rubber as spokes. Make it balance. Shine the infra-red light from the lamp on one side of the wheel, and it will turn.

REASON: The heat will expand the rubber bands, throwing the wheel out of balance.

The opposite is true: heat contracts rubber rather than expanding it. The wheel will turn the opposite way. (If the wheel is well balanced, see if bright sunlight will turn it. Place the wheel in the sun, covering half of it so the light shines only on the other half.)

A TRICK LIGHT BOX

NEEDED: A box, two lamps in sockets, two switches, glass, wood, candle, jar, lamp cord and plug.

EXPERIMENT: Build the box with lid. Paint the inside dull black. Arrange the glass and wood as shown. The author's box is 24 inches square by 7 1/2 inches deep, made of quarter-inch plywood.

Look through the peephole, and turn on one switch. The jar shows. Turn it off and turn on the other; the candle shows. Turn on both at the same time, and the candle is seen inside the jar.

REASON: The glass both reflects and transmits light. More light is transmitted when the bright object is beyond the glass. More is reflected when the bright object is on the same side of the glass as is the peep hole.

It is necessary to move the objects while both lights are on to mount them in proper position.

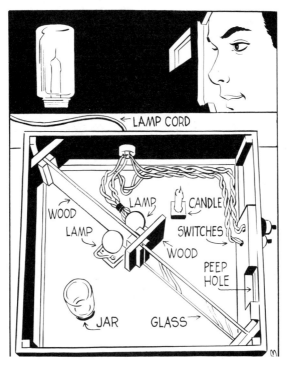

A Trick Light Box

A SIMPLE SHOCKER

NEEDED: A toy train transformer and two rods to hold.

EXPERIMENT: Turn the transformer on low. Squeeze the rods in the hands, and touch them to the transformer terminals. If no shock is felt, turn the transformer to a higher setting.

A Simple Shocker

REASON: The low voltage from the transformer cannot force enough current through the body to be dangerous, since the normal resistance of the human body is about 50,000 ohms. But do not use any other kind of transformer for this.

The author used carbon rods out of dry cells. They give a good contact when held in the hands. Large nails will do—or any kind of metal rods.

A STRONGER SHOCKER

NEEDED: A radio output transformer, Midland 15-611 or equivalent, a flashlight cell, tin can metal, wire, wood for a case (a cigar box will do), wire, screws or large-headed tacks.

EXPERIMENT: Connect one of the small wires to the cell. Connect the two larger wires to the tacks that will be used as terminals for the fingers. Connect the other small wire and the other end of the cell to the push switch which can be made of tin can metal.

Have someone touch the tack terminals lightly. Jiggle the switch up and down rapidly, and a shock is felt at the terminals. No shock is felt if the switch is held closed.

REASON: The transformer steps up the voltage only when the primary current is turned on and off quickly at the switch. For a weaker shocker, use a smaller transformer, such as Olson T-231.

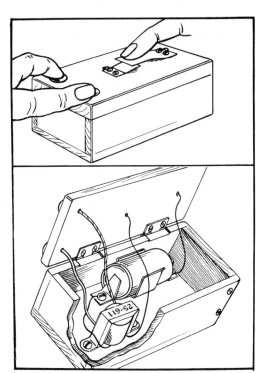

A Stronger Shocker

A SAFE SHOCKER

NEEDED: Three dry cells or four flashlight "D" cells, a small bell transformer, a bell or buzzer, a push button, some wires, a board for mounting the device, some furniture gliders or other metal pieces to serve as electrodes.

EXPERIMENT: Mount the parts on the board as shown. When the bell is ringing, terminals 1 and 2 will give a shock; terminals 3 and 4 will give a much stronger shock which may be felt through a line of 15 people holding hands.

REASON: Voltage from the cells is too low to give a shock, but the coils in the bell and the making and breaking of the circuit by the points in the bell can build up peak momentary voltages above 100. This is enough to give a shock.

The peaks, going through the bell transformer backwards as shown in the diagram, are increased 6–16 times, and this higher voltage is felt from terminals 3 and 4. Two or three cells may be used, to increase or decrease the shock.

All of these shockers are safe.

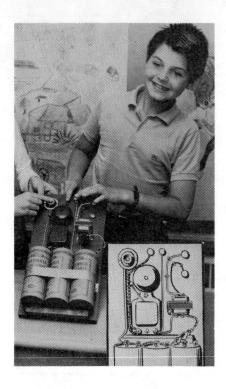

A Safe Shocker

BLINKY THE BLINKER

NEEDED: A battery of 90 or more volts (D), two neon lamps, NE51 or equivalent (A); one condenser of about one microfarad capacity (B); two resistors, 1/4 or 1/2 watt, one megohm each (C); wire for connections and an empty cigar box.

EXPERIMENT: Connect the parts as shown in the diagram. (If only one blinker light is wanted, use diagram at right with one resistor). The author made a cover for the box out of felt, leaving holes for the lamps to stick through to make the blinking eyes. This makes a good conversation piece, since the blinking continues for months, with no way to switch it off. Several bulbs may be connected to the same battery, using the diagram at the right, and they will blink without any regular sequence.

REASON: When the circuit is completed, one neon bulb fires before the other (it would be a very rare coincidence if both fired at once). While the first lamp is glowing the condenser charges through the resistance in series with the second lamp. The polarity of the condenser is such that as it builds up a charge the voltage across the first lamp is decreased to its extinguishing point. Then the second lamp begins to glow and the condenser begins to charge in the opposite direction through the other resistance. The voltage in the glowing lamp begins to decrease, until the lamp is extinguished, and the cycle continues to repeat over and over.

Blinky the Blinker

THE HILSCH TUBE MYSTERY

NEEDED: Vacuum cleaner hose that will blow air *out*, cardboard or wood box, two thermometers, a book, paper, glue.

EXPERIMENT: Make a hole in the middle of the box for the hose end. Place thermometers in both ends of the box. Glue the paper on one end, place the book at the other end to partially lock the air flow. Hold the hose at an angle as shown. Place the book so that some air will flow from the paper end of the box; the blowing paper will be the indicator.

Read the thermometers. Let the air blow five minutes, then read again. The temperature shown on the two thermometers will be slightly different.

REASON: Physicists can talk for hours on the principle that produces this effect, yet it is not completely understood. Well-made Hilsch tubes, running on compressed air, produce large differences in temperature. Such tubes are used in suits worn by workers in places contaminated by radioactivity to cool and keep them comfortable.

The author built the simple apparatus shown in the drawing. When the room temperature was 70 degrees he obtained readings of 93 degrees at one end and 99 at the other. The cleaner produced quite a bit of heat, warming the air that came from the hose.

The Hilsch Tube Mystery

A RHEOSTAT

NEEDED: A battery, flashlight bulb and socket, a long pencil (No. 2 or 3 lead), knife, tin can metal and shears to cut it, screws, tools, wire, wood.

EXPERIMENT: Assemble the rheostat as shown. When the lever is moved to the left along the graphite pencil lead, the light burns brighter.

REASON: The graphite offers resistance to the current, and the more of its length the current has to travel the more resistance there is. Resistance reduces the amount of current that can flow in the circuit. As is seen, the pencil must be split carefully to expose the graphite so the lever can touch it along its entire length. Do not break it. A six-volt battery and bulb probably will have to be used, since the graphite offers too much resistance for a smaller voltage.

Rheostats are used to control the sound volume in radios and televisions, and for other uses where variable amounts of current are needed.

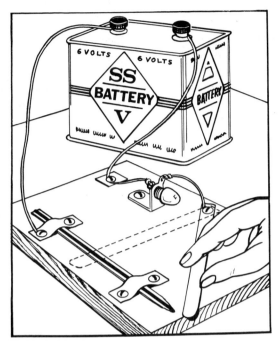

A Rheostat

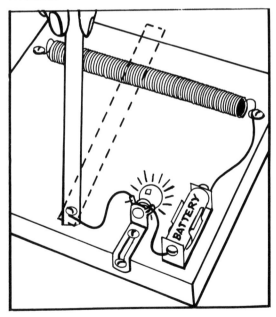

A Better Rheostat

A BETTER RHEOSTAT

NEEDED: A pen-light cell, a window shade spring, flashlight bulb, wood, screws, tin can metal, tools.

EXPERIMENT: Assemble the parts as shown, and when the lever is moved over the spring, the lamp glows brighter or dimmer.

REASON: The window shade spring is a conductor, but not the best. As the lever is moved so that more of the spring is in the circuit the lamp dims because a greater amount of spring offers a greater amount of resistance to the current.

The author, in building the model, placed a piece of half-inch dowel (from a lumber yard or hardware store) inside the spring to hold it sufficiently rigid.

It is good to clean the upper surface of the spring, where the lever makes contact, with fine sandpaper or steel wool. This will give better contact. The lever is made of tin can metal folded over to make it stiff enough.

Chapter 4
Tricks

A PEPPER TRICK

NEEDED: A glass of water, pepper in a shaker, soap.

EXPERIMENT: Sprinkle pepper on the surface of the water, draw the finger across it as shown in upper drawing, and the pepper will separate where the finger has been.

Challenge a friend to do the same. The pepper will close in behind the finger as shown in lower drawing.

REASON: Before drawing your finger across the water surface, secretly rub your finger on a bar of soap. The soap weakens the surface film of the water. Without the soap, the surface tension or film draws the pepper back to cover the surface of the water more rapidly.

A Pepper Trick

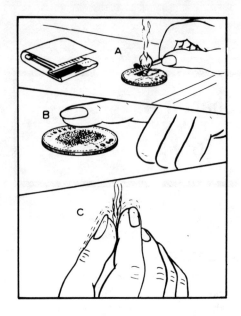

Smoke from the Figertips

SMOKE FROM THE FINGERTIPS

NEEDED: A book of safety matches and a half dollar.

EXPERIMENT: Tear off a thin piece of the striking surface as shown at A, place it on the coin, and burn it with a match as shown.

Rub the finger on the darkened place on the coin, as in B. Then, when the fingers are rubbed together as in C, a mysterious smoke will rise from them.

REASON: The striking surface of a safety match book contains free red phosphorus. The heat of the hands and the heat of friction from the rubbing causes a very slight union of phosphorus with oxygen of the air to form a white vapor of phosphorous oxide.

A HOLE IN THE HAND

NEEDED: A cardboard mailing tube or a tube made of paper.

EXPERIMENT: Look through the tube at a distant object, placing the tube at the left eye. Bring the right hand up beside the tube as shown. A hole will be seen in the hand.

REASON: The eyes see two images, but these are combined in the brain so that we are not confused. In this case we are somewhat confused; the right eye sees the hand, and the left eye sees the distant object, and these are not compatible in the brain.

It is an optical illusion.

A GYPSY MYSTERY

NEEDED: A glass of water and a cloth.

EXPERIMENT: Place the cloth (a cotton handkerchief will do) over the glass of water, turn it upside down, and the water will not run out. Have someone ask a question, and if the answer is "yes" the water will begin to "boil" when the finger is placed on the glass as shown in 2.

REASON: When the cloth is placed over the glass, be sure it is pushed downward, as shown in Figure 1. Hold the cloth when the glass is inverted, and when the finger presses down on the glass, let the glass slip downward into the cloth if you want the answer to be "yes."

Of course, the water does not actually boil, but seems to as the cloth is pulled tighter, because air bubbles will come through the cloth and float to the surface of the water.

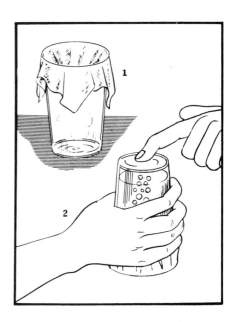

A Gypsy Mystery

MOVING MATCHES

NEEDED: Five wooden matches, a drop of water.

EXPERIMENT: Break each match in the center and arrange them as shown, so that the broken places touch. Let a drop of water fall so that all the breaks are touched by the water. The matches should not be broken completely in two parts.

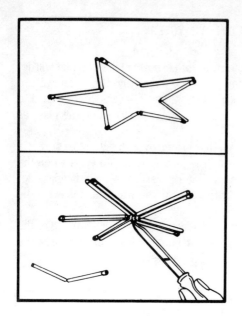

Moving Matches

Slowly, the matches will begin to move until they have formed a star.

REASON: This is an example of capillary action. The water moves by capillary action into the dry wood, swelling the cells. The swelling tends to make the matches straighten out somewhat.

IT'S MAGIC!

NEEDED: A polyethylene bag, a sharp pencil.

EXPERIMENT: Blow up the bag like a balloon, and twist the opening to make it air tight. Push a sharp pencil through the bag, in one side and out the other, leaving ends protruding.

The balloon does not burst and the air does not leak out.

REASON: Polyethylene plastic will seal itself around the pencil because of its peculiar molecular structure.

The thick polyethylene bags suitable for this experiment are found in grocery stores. Raw carrots are sold in them, and sometimes other vegetables. They are also used for wrapping frozen foods.

CAUTION: Keep all plastic bags away from small children.

SECRET WRITING

NEEDED: A porcelain table top or sheet of glass, two sheets of paper, some water, a hard lead pencil with a smooth rounded point.

EXPERIMENT: Wet one sheet of paper, place it on the smooth surface, put the dry paper over the wet one, and write with firm pressure on the dry paper.

The writing can be seen on the wet paper, but will disappear when the paper is dry. It reappears when the paper is wet again.

REASON: The pressure of the pencil on the paper compresses the wet fibers so that they reflect light in a slightly different way when wet. The writing will be dim, like the watermark on some stationery.

A PAPER TRICK

NEEDED: A card, a rule, a pencil, a sharp knife or razor blade.

EXPERIMENT: Cut the card as shown in the drawing at upper left. The card can then be opened out, so that the hole in it is large enough for a person to step through.

A piece of strong paper the size of a postal card, 3 1/4 × 5 1/2 inches, or even much smaller, may be used, in place of the card.

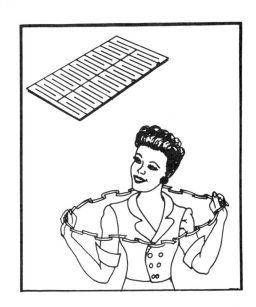

A Paper Trick

JUMPING BEANS

NEEDED: A capsule from a drugstore and a shot or small ball bearing.

EXPERIMENT: Put the ball into the capsule, place it on the hand, and as the hand is moved, the capsule flips over and over in a lively manner.

REASON: The mass of the ball times the velocity it acquires in free motion when the capsule is tipped to slope somewhat downward is called its momentum. The weight of the light hollow capsule is not sufficient to stop the ball, so it flips over as the ball reaches the rounded end.

Real "jumping beans" are entirely different. They jump because of the movement of a tiny live animal inside.

HINDU MAGIC

NEEDED: A glass jar with a narrow opening, a table knife, enough uncooked rice to fill the jar.

EXPERIMENT: Fill the jar with rice, and plunge the knife into it a few times. Then announce that you will cause the jar to rise. Now plunge the knife deep into the rice, raise it slowly, and the entire jar will be lifted.

REASON: Strangely enough, there is no hidden secret, although the trick has been performed by Hindu fakirs and called magic. As the knife is stuck a dozen times into the rice, it packs the rice more and more tightly. When the blade is finally plunged deep into the rice, the grains have been packed tightly enough to press against the blade with enough frictional force to lift the jar.

FIREWORKS FROM A LEMON

NEEDED: A candle flame and a lemon.

EXPERIMENT 1: Squeeze the lemon peel near the flame as shown in the drawing, and small displays of "fireworks" may be seen shooting from the flame.

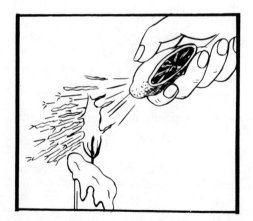

Fireworks from a Lemon

REASON: As the lemon peel is bent, some of the oil and water in it squirt out into the flame. Some of the oil burns as it passes through the flame, and some of the water vaporizes and sputters.

EXPERIMENT 2: Sprinkle flour on the candle flame. Tiny sparkles will be seen as the flour particles catch fire. The particles must be fine, with a large proportion of their surfaces exposed to oxygen of the air, to produce this effect.

FLIP THE PENNY

NEEDED: A one-cent piece.

EXPERIMENT: Place the coin as shown in the upper photo, on the knuckle of the little finger. Flip the end of the finger with the thumb as shown. After a little practice, the coin can be made to turn over.

REASON: The upward movement of the finger simply gives the coin a twirling toss into the air. The little finger is hinged just beyond the coin and the sudden rotary motion of the finger gives rotary motion to the coin. It is necessary to flip the finger downward with the thumb. The little finger must be on the thumbnail. If the positions of the thumb and finger are reversed, the trick will not work.

The scientific principles involved here include muscle elasticity, inertia, momentum, and gravity.

THE MAGIC PENCIL

EXPERIMENT: Move your finger toward a pencil and make the pencil roll away mysteriously without touching it.

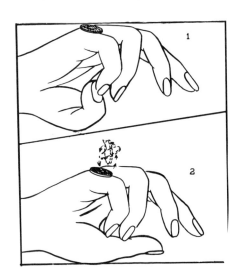

Flip the Penny

REASON: This is a trick that employs a scientific principle or two. As the finger approaches the pencil, blow on it, and the breeze will roll the pencil away.

The force exerted by the air as it moves to the pencil makes the pencil move, while a psychological factor magicians call "misdirection" comes into play, making the onlooker try to connect the finger movement with the pencil motion, when it actually has nothing to do with it.

FAST MONEY

NEEDED: A dollar bill, preferably a crisp one.

EXPERIMENT: Hold the dollar bill as in drawing at upper left. Allow someone to hold his hand ready to catch it. As it is dropped it will move between the thumb and fingers before the lower hand can grasp it.

REASON: It takes time for the eye to see that the bill has been released, time for the brain to tell the hand to grasp, and still more time for the muscles in the hand to obey.

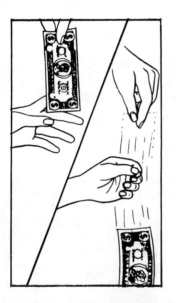

Fast Money

A MAGNETIC MYSTERY

NEEDED: A magnet, paper, an iron paper clip, string, book, pencil.

EXPERIMENT: Wrap the magnet so that it is hidden; arrange the string and clip so that the book and pencil hold them down, and you have a mystery trick.

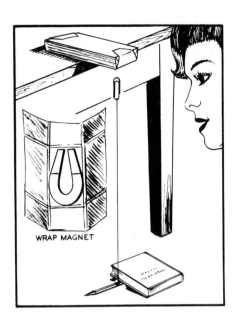

A Magnetic Mystery

WRAP MAGNET

OBSERVATION: The magnet pulls against the iron in the paper clip, holding it up. The clip will stay suspended until it is moved far enough way from the magnet for gravity to exert a stronger pull than the magnet.

A BALLOON TRICK

NEEDED: A small rubber balloon, a drinking glass, some water.

EXPERIMENT: Hold the balloon in the glass as you blow it up. It expands against the glass and holds so tightly that the glass of water can be lifted by it.

REASON: The friction of the rubber against the glass may be sufficient to prevent its pulling out easily. But if the friction is not great enough, the balloon still will not pull out easily because air cannot get past it into the bottom of the glass. As it is pulled, the air pressure in the glass is reduced so that the greater pressure of the atmosphere on the bottom of the glass, exerted upward, can lift the weight of the glass of water.

THE SALT SHAKER MYSTERY

NEEDED: A salt shaker with salt and a steady table top.

EXPERIMENT: Place the shaker on a small pile of salt and balance it. Blow the salt away, and the shaker can remain balanced.

REASON: You do not blow *all* the salt away. The few grains needed to balance the shaker are held by it so that they do not blow away, although they may not be seen without very close examination.

This will require a little practice. If you blow too hard, the balance may be upset.

A LEAF PRINT

NEEDED: A warm crayon, a leaf, paper, a hot iron.

EXPERIMENT: Rub the back side of the leaf with the wax crayon until it is well covered with the wax. Place it between two sheets of paper, and press with a hot iron. The wax melts off of the leaf, making a pattern on the paper.

No explanation is necessary in this experiment, except that it is not necessary to place the hot iron on the leaf, since thin paper conducts the heat rapidly to the leaf, and the paper cannot smear the iron with any undesirable substance from the leaf.

A Leaf Print

The Ouija Board

THE OUIJA BOARD

NEEDED: A ouija board and a partner.

EXPERIMENT: Place fingers lightly on the movable planchette. Do not rest arms or elbows on the table. Ask the ouija a question and the planchette may move to spell out the answer.

REASON: The brain and the subconscious mind control the movements of the arm and finger muscles involuntarily. The answers are usually those the players want them to be.

Many people take the ouija seriously, believing it is a link to the spirit world. But remember, if the spirits were moving the planchette, it would move without the fingers on it. Play ouija as a game only.

AN OPTICAL ILLUSION

NEEDED: Cardboard, scissors, protractor, pencil.

EXPERIMENT: Mark the card with the protractor, and cut it as shown. Write X on one piece, and O on the other. Place them so the O is above the X, and the X seems larger. Reverse the positions, and the O seems larger.

REASON: This is just another of the many optical illusions that may be made with simple materials. The brain takes into consideration not only what the eye sees at the moment, but the memories of what the eye saw before, and the meanings of those things the eye has seen before.

This is good; it enables us to observe more accurately, but sometimes it creates illusions, too.

An Optical Illusion

DO YOU DARE?

NEEDED: A two-pound coffee can of sand, suspended overhead on a string, a "victim."

EXPERIMENT: Have the "victim" stand against the wall, pull the heavy can back until it touches his nose, then let it go (don't push.) See if he can stand still as the can swings back toward his nose.

REASON: According to the law of conservation of energy the can will never come back far enough to hit the "victim" on the nose. A little energy is always lost as the can swings. Of course, if the can is pushed instead of merely being released, it can come back far enough to hit the nose.

The author weighted the can with a brick instead of sand.

Do You Dare?

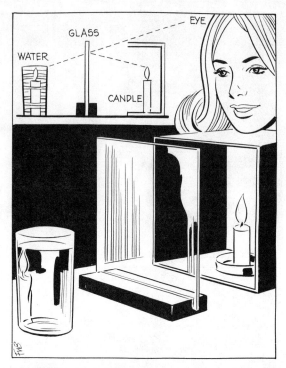

A Candle Burning in Water

A CANDLE BURNING IN WATER

NEEDED: Window glass, a glass of water, a burning candle in a can or box.

EXPERIMENT: Mount the glass in a slot in a board so it stands upright. Place the candle in the can in front of the glass as shown, so its reflection is seen in the glass. Stand the glass of water behind the window glass, place it correctly, and the candle will seem to be burning under the water.

REASON: The window glass reflects some light and lets some light pass through it. Light from the glass of water passes through to the eye, and at the same time some of the light from the candle is reflected from the window glass to the eye, creating the illusion. The drawing shows the paths of the light rays.

A TRICKY PUZZLE

NEEDED: A glass of water, a magnet, tacks and small nails. Put the tacks and nails into the water.

A Tricky Puzzle

EXPERIMENT: Challenge someone to get the metal out of the glass of water without pouring out the water and without reaching into the water.

Hold the glass up, and place the magnet opposite the tacks and nails, touching the glass. The metal will be attracted to the magnet, and may be made to slide up the side of the glass. As it reaches the top of the glass it will cling to the magnet and can be moved away.

This cannot be done easily with a tin can used instead of the glass. The magnetic lines go through glass, or aluminum, or any other non-magnetic material, but not so readily through the metal of the tin can. This is because a "tin" can is only about two percent tin. The rest is steel, which is magnetic.

THE RISING BALL

NEEDED: A ball, a glass jar, a table top.

EXPERIMENT: Place the jar over the ball, and move the jar with a rapid rotating motion until the ball leaves the table top and ascends up into the jar. Keep the jar rotating, and it may be lifted from the table with the ball inside.

The Rising Ball

REASON: Centrifugal force may be defined as a force that tends to make rotating bodies move away from the center of rotation. As the ball moves away from that center it climbs up the rim of the jar and remains there as long as the jar is rotated rapidly in a circle.

The ball must be large enough that its center as it rests on the table is higher than the rim of the glass jar. Then it can climb up the rim. A large light-weight ball such as a tennis ball is good for this experiment.

An Impossible Trick

AN IMPOSSIBLE TRICK

NEEDED: Writing paper and scissors.

EXPERIMENT: Cut the paper as shown, hold it by the outer edges, pull or jerk, trying to make it tear apart at both cuts at the same time. It is impossible in almost every try.

REASON: No matter how carefully the cuts are made, they are never quite the same. The pull at the cuts is never quite the same. And the paper itself cannot be exactly alike at both cuts. For these reasons the paper does not tear alike at both cuts. It is more likely to tear at both cuts when a quick jerk rather than a steady pull is applied to the paper.

The trick can be done sometimes if the middle section of paper is weighted so that its inertia tends to keep it in the same place as the outer sections are torn away.

MAGIC PAPER

NEEDED: Tissue paper, a gummed envelope flap, a dish, a match.

EXPERIMENT: Roll the tissue paper and stick the edges with tiny pieces of the gummed flap of the envelope. Stand it on the dish, set fire to the top, and as the flame burns toward the bottom the remains of the tube will rise into the air.

REASON: The heat of the flame produces an upward current of air—enough to lift the light-weight paper and ash.

The idea for this came from the magicians' magazine *The Linking Ring*. Magician Bill Pitts tells in the magazine how he

Magic Paper

extracts a mysteriously vanished dollar bill from the burning tube. That is magic; the science part is the rising of the burning tube.

The paper rolled into a tube may be stuck with tiny bits of rubber cement instead of paper. Try different size rolls and different paper. Magician Pitts recommended a roll two inches in diameter and eight inches high.

As always, be careful of fire. It is always dangerous.

WHICH ONE'S WHICH?

NEEDED: Two people.

EXPERIMENT: Have someone cross his hands and hold out his fingers. Point to one finger and ask him to move it quickly. Often he will move the wrong finger.

REASON: The eyes and the brain do not work together in the normal pattern when the hands are crossed. We respond in a manner to which we have been trained. We can learn new behavior, but it is a slow process requiring considerable practice.

With practice the right finger can be moved when pointed to.

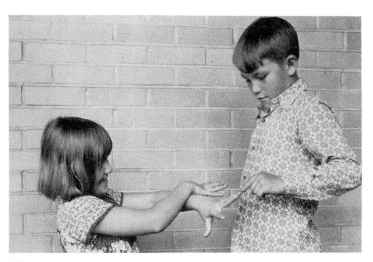

Which One's Which?

A COOL BOOK

NEEDED: A book with a slick-paper dust jacket.

EXPERIMENT: Hold the book as shown, and let it slide quickly through the hands until it comes to rest on the knees. It will feel cool as it comes to rest between the hands.

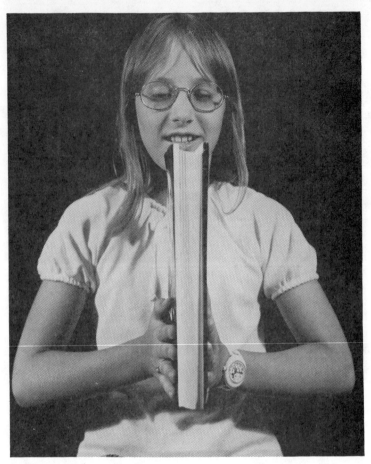

Cool a Book

REASON: This was in the March, 1975, issue of the magazine *The Physics Teacher*. It was discovered by Dr. Robert Everett Vermillion, of the University of North Carolina at Charlotte, and was presented with the invitation to readers to explain it.

It is this author's suggestion that, as the book comes to rest, the hands press more tightly against it, allowing heat to travel more readily from the hands to the book.

PICTURE TRANSFER

NEEDED: Colored comics from the newspaper, turpentine, paper towels, detergent, water, a spoon, a dish, school tablet paper, a jar.

EXPERIMENT: Put one ounce of detergent, two ounces of turpentine, and four ounces of water into the jar, tighten the lid, and shake until an emulsion is formed. It will be white. Dip a newspaper picture in the solution, blot it lightly with paper towel, then lay it face down on the paper. Rub it thoroughly with the bowl of the spoon, lift it off, and the picture will be seen in reverse on the tablet paper.

REASON: Printer's ink is dissolved by the turpentine, and so is transferred to the other paper. The purpose of the detergent is to make the turpentine droplets remain uncombined to form a uniform emulsion.

Picture Transfer

THE MATCH BOX DROP

NEEDED: An ordinary small box of matches.

EXPERIMENT: Open the box half an inch, and hold it in the hand so a friend cannot see that it is open. Hold it a foot above the table and drop it. It will stand on end. When the friend tries it, it falls over.

REASON: Of course, the friend tries the trick with the box closed.

When it is dropped open the closing of the box as it hits the table slows the downward motion, allowing the box to settle more easily and not bounce.

The Match Box Drop

THE STRAW-THROUGH-POTATO-TRICK (1)

NEEDED: Soda straws, raw potatoes, a needle.

EXPERIMENT: Unless the potatoes are fresh, they should be soaked in water to soften the peel. Hold the straw firmly between thumb and finger, hit the potato quickly with the end of the straw. With a little practice the straw may be made to go all the way through the potato.

REASON: The straw is a very strong structural shape, and, for this reason, if it is straight, it may be forced through the potato. The myth persists that air trapped in the straw makes it strong enough to penetrate the potato.

Now, make needle holes in the straw, being careful not to bend or crush it. It will still penetrate as well, sometimes easier, thus disproving the myth. If, after practice and an hour's soaking of the potatoes, this does not work easily, use another type of soda straw.

ANOTHER STRAW-THROUGH-POTATO TRICK (2)

NEEDED: Straws with a flexible section, tin can metal, tin snips, potatoes.

EXPERIMENT: Cut a piece of the tin can metal so that it slides over the straw through the grooves as shown. Then, by pushing against the metal the straw may be thrust through the raw potato without closing the passage through the straw at all.

REASON: This is another proof that the air trapped in the straw has nothing to do with the experiment. The straw can penetrate the potato simply because it is a tube, and tubes are relatively strong if force is exerted along their length.

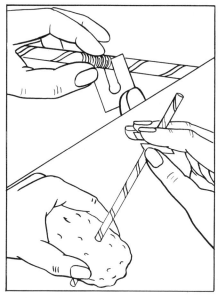

The Straw-Through-Potato Trick (2)

INVISIBLE WRITING

NEEDED: Salt, water, writing paper, carbon paper.

EXPERIMENT: Use as much salt in quarter of a glass of water as will dissolve. Let it set until the undissolved crystals have settled. Write with a toothpick, using the salt solution.

When the writing is dry it is invisible, or practically so. But the salt water has left rough crystals on the paper. When the carbon paper is rubbed over the writing paper the writing stands out and can be easily read.

REASON: The salt crystals have rubbed off more of the carbon than the smooth paper.

LIGHT FROM THE TEETH

NEEDED: A totally dark room, a mirror, wintergreen Life Saver candy, pliers.

EXPERIMENT: Stay in a dark room until the eyes are adapted to the darkness (usually ten to fifteen minutes). Look into the mirror, and, with lips apart, bite through the mint. A flash of bluish light is seen as the candy is crushed.

REASON: Certain solids when crushed or fractured produce a difference of electrical charge on the pieces. Wintergreen oil absorbed on sugar is one of them. The electricity discharges across the surface of the solid to equalize the charges, exciting molecules or atoms of the solid or the air at its surface, thus producing the light. The small amount of electricity produced cannot be felt. But to avoid any possibility of breaking a tooth the mint may be crushed with pliers to produce the light flash.

THE EDIBLE CANDLE

NEEDED: An apple, almond nut slivers, a knife, a match.

EXPERIMENT: Cut a piece of apple to look like a candle. Place a sliver of almond in the top to make a wick. Light it, and it will burn like a candle.

REASON: Oil in the nut meat will burn for a few seconds like a candle wick. The apple does not burn and may be eaten.

A CHEMICAL TRICK

NEEDED: Cornstarch, tincture of iodine, a glass of water, pan for boiling.

EXPERIMENT: Put three drops of the household iodine into the glass of water. It should turn slightly brown. Have a friend stir it with a finger; nothing happens. Now stir it with your finger; and the color changes to blue.

REASON: Prior to the demonstration, boil half a teaspoonful of cornstarch in enough water to fill a glass half-full. Let it cool. Before the demonstration, dip a finger into the starch water. Then when that finger is used to stir the iodine water the familiar test for starch takes place—the blue color appears. Iodine and starch combine to form a somewhat mysterious temporary combination called "starch iodide," which is blue.

WEDGE MAGIC

NEEDED: Vase or detergent jug, soft rope, a rubber ball.

EXPERIMENT: Blackstone, the magician, used a two-foot rope, placing one end of it into the vase, winding the rest around the neck of the vase. He turned the vase over, letting the rope fall free. It did not fall out of the vase. Then, holding the rope, he turned the

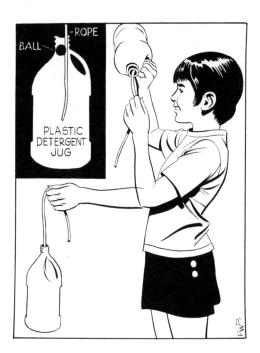

Wedge Magic

whole thing over, and held the vase suspended by the rope. Then he handed out both vase and rope for inspection.

REASON: Inside the vase is a small rubber ball. Its diameter plus that of the rope must be slightly greater than the diameter of the neck of the vase. When the vase is upturned, the ball wedges itself against the rope in the neck and holds tightly. Turn the vase upright again, push the rope in slightly to release the ball, pull the rope out, and hand it to the spectator for examination. Take the vase by the neck, and in handing it to the spectator for examination, invert it so the ball falls out into the hand where it can be concealed.

SYMPATHETIC VIBRATIONS

NEEDED: Two drinking glasses (goblets are best), some vinegar, water, a fine wire.

EXPERIMENT: Put water into the glasses, put a little vinegar into each glass, and dip a finger into the vinegar water. Rub the finger around the rim of each glass, making a musical tone. Put in or take out water from the second glass until the pitch is exactly the same in both glasses. Glasses with stems are easier to hold and vibrate better. Place the wire across the rim of the second glass, and as a musical tone is produced in the first glass the wire will be seen to move slightly on the rim of the second.

REASON: The sound from the first glass sets up resonant vibrations in the second because the two glasses have the same frequency of vibration. This idea is used in musical instruments.

If good thin glasses are used and they will vibrate well, try producing a tone in the first glass, then putting a hand over it to stop the sound quickly. The sound should then be heard coming from the second glass.

A GRAVITY TRICK

NEEDED: Thin plywood or thick cardboard, a knife, a leather belt.

EXPERIMENT: Cut the plywood in the shape shown. When the belt is placed in the slot, the wood with the belt can be held on the finger, seeming at first glance to defy gravity.

REASON: The belt pulls down on the piece of wood. As the large end of the wood starts to move down, the ends of the belt move closer to the body than the supporting finger. The belt and wood may be held at rest in this position because the center of gravity is directly under the finger. This trick will not work with an ordinary weight, but requires a leather belt which is stiff enough not to bend easily.

A Gravity Trick

BERNOULLI WITH A DIME

NEEDED: A dime and a smooth table top.

EXPERIMENT: Challenge a friend to place the dime on the table, then, without using his hands or arms, turn it over.

REASON: Blow hard over the top of the coin. Bernoulli learned that air in motion has less lateral pressure than air at rest. The air moving over the coin has less pressure than the air below it, and so the air below flips it over. The coin cannot be turned over by blowing air down on it.

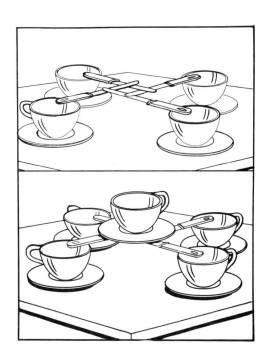

A Bridge of Knives

A BRIDGE OF KNIVES

NEEDED: Four table knives and five objects, (cups will do).

EXPERIMENT: Build the bridge of knives as shown, and the fifth cup can be supported on the knife blades.

REASON: Each knife is supported at both its ends, although to some it would not seem so at first glance. When the extra weight is added each knife bends downward slightly until the strain is proportional to the stress. The upward and downward forces are then equal, and friction prevents the knives from sliding sideways.

Blow the Coin Over

BLOW THE COIN OVER

NEEDED: A funnel and a coin.

EXPERIMENT: Hold the funnel upright, with the coin inside. Try to turn the coin over by blowing hard into the funnel as shown. It is difficult if not impossible. Cover the spout with a finger and the coin may be blown over.

REASON: With the spout uncovered, some of the air passes through the small spaces between the coin and the side of the funnel, tending to equalize the pressure below and above the coin. When the spout is covered air cannot get out below the coin, and more is blown across the top of the coin. According to the Bernoulli principle, the moving air above the coin has less pressure than the still air below, and tends to lift the coin slightly. Then the air catching below it flips it over.

A SALT TRICK

NEEDED: Soft blotter paper, salt, water, matches.

EXPERIMENT: Heat the water, and dissolve in it all the salt it will take. Soak some strips of blotter paper in the solution, dry them, then light one and let it burn half way. After it cools place matches on it as shown and see how many it will hold. An untreated blotter burned the same way falls into ashes and will hold no weight.

REASON: The heat of the burning paper fuses or melts some of the salt crystals together. When they cool they provide the strength to hold the matches. The author, in trying this, succeeded in placing 44 kitchen matches on a two-inch strip of burned blotter before it broke under the weight.

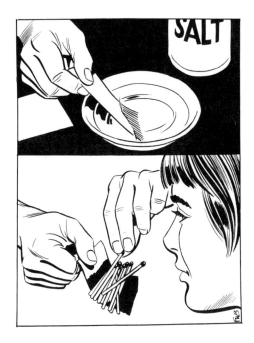

A Salt Trick

A FORK AND SPOON TRICK

NEEDED: A glass, a fork, a spoon, two matches.

EXPERIMENT 1: Balance the fork, spoon and match as shown. Set fire to the ends of the match. The wood will burn to the edges of the fork and glass, but the fork and spoon will not fall.

REASON: The wood on the match must be supplied with enough heat to make it burn. When the flame reaches the glass and the fork, they conduct so much of the heat of the flame away that the wood does not burn farther. Enough of the solid wood remains to hold the weight.

EXPERIMENT 2: Lay a match and a cigarette side by side on a piece of board. Light both. The wooden match will not burn past

A Fork and Spoon Trick

the edge of the wood; the cigarette will burn all the way. The cigarette contains a chemical that promotes burning, to prevent its going out while in the hand. This is good for the smoker, but makes the cigarette more dangerous as a fire hazard.

Chapter 5

Biology & Psychology

CAN YOU HOLD A PENCIL?

NEEDED: A pencil.

EXPERIMENT: Hold the pencil (or other very light object even a feather will do) at arm's length for 20 minutes. You'll likely find it impossible to hold it even five minutes.

REASON: While the muscles of the arm are strong enough to lift and hold a much heavier object, they have not been trained and strengthened to support the weight of the hand and arm for a period of several minutes.

This, like many other physical feats, may be performed after considerable training and practice. A few *very strong* people may do this 20 minutes or more the first time they try it.

GHOST LIGHT

NEEDED: A bare electric bulb in a dark room.

EXPERIMENT: Turn on the light, gaze at it for ten seconds, then switch it off. You will continue to "see" light.

REASON: While the workings of the eye have not been satisfactorily explained, we know that the eye continues to see an object for a short period after the object is actually no longer visible.

This is called "persistence of vision." It is what allows us to see constant moving pictures at motion picture theaters or on our television screens, although these pictures are constantly flickering off and on.

SENSE OF TOUCH

NEEDED: Two pencils and a partner.

EXPERIMENT: Touch the two pencils to the finger as shown, and the partner can feel both points easily. Try touching the partner on other parts of the body, and it will be difficult to tell whether one or two pencils are used.

REASON: Nerve endings that give us our sense of touch are close together in the fingertips but farther apart in most other parts of the body. Try the cheek, the tongue, the back, the arm, the leg, and the back of the hand.

It will be interesting to touch the pencil points close together, then repeat at the same place on the body with the points farther apart. See how far they must be in order that the partner can feel two separate pencils.

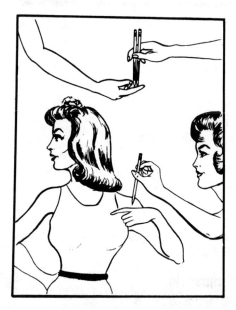

Sense of Touch

TWO LEAVES ALIKE

NEEDED: Leaves and marbles.

EXPERIMENT: Try to find two leaves exactly alike. It is impossible.

REASON: Nothing visible in living nature is exactly like anything else. In leaves, even if the tops seem alike, differences in the vein patterns may be seen. Crystal patterns of the same substances may be identical as well as molecules and smaller particles which we consider here as invisible.

To illustrate how impossible it is that collections of molecules or cells in living organisms could have the same configuration, place marbles together on the living room rug, and push another marble into the collection. Many of them will move; and it would be impossible to roll a marble against one group a second time and make the members move in the same way and form the same pattern.

WATER FROM LEAVES

NEEDED: A live plant, a sheet of cellophane.

EXPERIMENT: Wrap the plant with cellophane, tying it tightly around the stem. Droplets of moisture will form inside the cellophane.

REASON: A plant takes in water through its roots and gives off water through the leaves, flowers, and stems but mostly through the leaves. Not all the water taken in is given off; some of it is used by the plant in the manufacture of its food.

A potted plant or one growing in the ground is better, but the experiment may be performed with a stem cut and placed in water, as shown here.

MUSCLE HABIT

EXPERIMENT: Try patting your head with both hands at the same time, and it is easy. Try rubbing your stomach with one hand and patting your head with the other, and it is quite difficult.

REASON: Automatic back-and-forth motions or circular motions of the hands and arms are performed easily, because the nervous system has been adapted to them through much repetition. Any change from the pattern formed is difficult and must be controlled by active thought in the brain. If your power of concentration is sufficiently strong, it can overcome the habit pattern and these motions can be performed.

WHY GET DIZZY?

NEEDED: Only yourself.

EXPERIMENT: Turn around rapidly, on foot or on a revolving stool, and soon you are dizzy.

REASON: Why? The cause is not completely understood. It is thought, however, that fluid in the semicircular canals of the ear begins to move around as we turn, tiny calcium compound particles in it brushing against tiny projections and registering the turning motion in the brain. When we stop turning, the fluid continues to turn, and the brain interprets this as if we are still in motion our-

Why Get Dizzy?

selves. The motion of the fluid soon stops after we stop, and we are normal again.

HOT OR COLD?

NEEDED: A friend, a match, an ice cube.

EXPERIMENT 1 : Have the friend cross the fingers of one hand, behind the back. Touch the crossed fingers with one finger, and the friend will believe two fingers were used.

Hot or Cold?

EXPERIMENT 2: Tell the friend "I will blow out the match flame, and will touch you with the match. See if it still feels hot to you." Instead, touch the friend with the ice cube. The friend is likely to believe heat, not cold, is felt.

REASON: Our sense of touch operates very much by habit. Our fingers are not accustomed to the crossed position, and the touch of one finger registers as two.

Our nerves that sense heat and cold can be confused, and, if we are expecting heat, the cold ice cube may register in the brain as heat.

THE RISING ARMS

NEEDED: A doorway.

EXPERIMENT: Stand in the doorway, press outward with the hands and arms as shown in the drawing at right, and count slowly to 25. Step away from the door, and the arms begin to rise mysteriously.

REASON: This is an interesting example of the workings of mind and muscle. The 25-count effort is sufficient to produce a persistent attempt to raise the arms. The doorcase prevents this, but as soon as you step out of the doorway, the persistent effort to raise the arms becomes a possible reality.

The Rising Arms

A TRICK OF THE MIND

NEEDED: Two opaque jars or cans and enough heavy metal to fill them.

EXPERIMENT 1: Fill one of the jars with something heavy, and screw on the lid so that the contents cannot be seen. Both jars should look alike. Have someone approach the table quickly and pick up the jars, one in either hand. The empty one will go up; the heavy one will be hardly raised from the table. (The person performing the experiment should not know that one is heavy.)

EXPERIMENT 2: If there is access to a chemical lab, find a container of mercury and ask someone to hand it to you. There may be great surprise when the friend finds he does not lift the container on the first try. The mind simply is not prepared to find it so heavy in proportion to its size.

REASON: Our minds direct the actions of our muscles, usually without our thinking about it. The mind's eye sees the two similar jars, and directs the muscle power it should take to raise them up, but, since they look alike, the muscles in both arms are commanded to lift equal weights. It takes an interval of time for the command to the muscles to be made when the mind learns that the jars are not equally heavy.

BREAD MOLD

NEEDED: Glass jar or dish, foil, wire, water, a piece of bread.

EXPERIMENT 1: Hang the bread on the wire inside the jar, as shown, or simply place it on a dish. Put water in the jar or dish to keep the bread moist. The jar may be covered with foil to keep the moisture in. Mold will eventually appear on the bread.

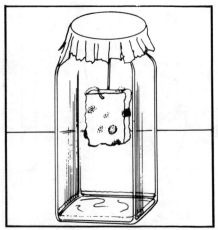

Bread Mold

EXPERIMENT 2: Examine some of the growths under a microscope or magnifying glass. Note their interesting shapes.

REASON: Fungus spores, including those that produce bread mold, are always in the air. They fall on the bread, and, when conditions are right, they grow into mold. It may take several days for a good covering of mold to appear.

The mold will be in several colors. A blue-green kind which is likely to appear is penicillium, a member of the genus from which penicillin is extracted.

A MOLD-CULTURE MEDIUM

NEEDED: Potatoes, gelatin, a cooking pan, a large dish.

EXPERIMENT 1: Boil the potatoes until they disintegrate. Add one-fourth as much gelatin by volume, and pour into the dish to jell.

EXPERIMENT 2: Set a dish of the culture in a vessel of water and boil the water five minutes. Cover the culture with plastic and set it aside. Heat will probably have killed the tiny plants, and none will grow, at least for some time.

OBSERVATION: If the dish is allowed to set overnight at room temperature, mold will begin to grow. There will be spots in several colors, and the whole may become very attractive as the various kinds of molds spread to cover the dish. Leave the dish uncovered for this experiment.

REASON: The air always contains mold spores, as well as other micro-organisms. The spores settle on the culture medium and begin to grow. Examine their structure under a magnifying glass. They are plants, but do not manufacture their food. They must get their food from the culture, or from living or decaying animal or vegetable matter.

GROW A FUNGUS GARDEN

NEEDED: A dish of culture medium, a piece of molded bread, a wire, a source of heat.

EXPERIMENT: Heat the wire to kill spores that may be clinging to it. Let the wire cool, and use it to transfer various colors of molds in any pattern desired. Heat and cool the wire each time a different kind of mold is touched with it. Cover the medium to keep out unwanted spores.

Try transplanting molds from blue cheese and other substances. Be sure all pieces of glassware are sterile before use. A handle on the wire will keep it from burning the fingers.

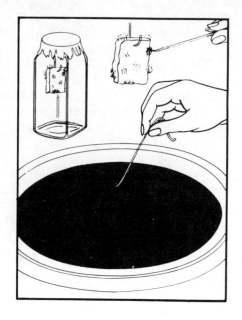

Grow a Fungus Garden

OBSERVATION: The culture medium will be covered in a few days with a beautiful "garden" of vari-colored molds. Molds are forms of fungi, plants that manufacture no food for themselves, but live as parasites.

(Experiment suggested by Al DeAnda, of the Mycology Lab, White Sands Missile Range.)

Tree Force

TREE FORCE

NEEDED: Observation.

The roots of a growing tree can exert unbelievable force. They can destroy concrete foundations and create massive upheavals of earth and rock. They can split a hard rock if growing inside it.

In the drawing a tree growing in a front yard is seen destroying concrete steps, the walk, and the street curb.

The large force exhibited by the roots is the combined forces of millions of tiny cells with fragile walls.

But there is very little energy involved. Energy or work equals force times distance moved. Power equals energy over time. Distance moved is small, energy small, time long (large), power small.

AN ILLUSION

NEEDED: This drawing.

EXPERIMENT: Make short, swift, motions with the drawing of circles, with the center of the circles moving in a small circle. The drawing will seem to turn like a wheel.

REASON: The apparent rotation of the circles is due to persistence of vision. This means that the eye sees the image after the visual stimulus ends, that is, after the circles have moved to another point.

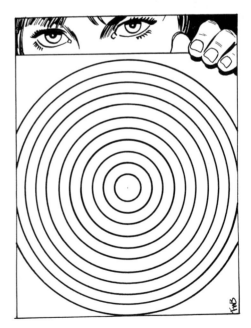

An Illusion

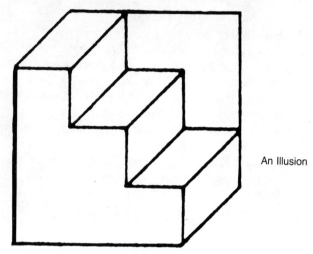

An Illusion

This persistence of vision is useful to us when watching a movie or television. The pictures there do not remain constant; they go off and on. Yet our eyes do not turn them off when they actually go off, and so we get the impression of a constant but moving picture.

AN ILLUSION

NEEDED: The drawing shown here or a larger one like it.

EXPERIMENT: Look at the drawing of the staircase. Is it right side up or upside down? Look again and it likely will be reversed.

COMMENT: This is an old illusion known as the Schroder Staircase. It is called an equivocal figure and illustrates fluctuations of the vision process.

AN INDOOR GREENHOUSE

NEEDED: A pan or shallow box, coat hanger wire, plastic bag.

EXPERIMENT 1: Bend the wire so it makes a loop over the box as shown. Pull a plastic bag over it and tie the ends. Pull the excess under and fasten it with tape. This makes a greenhouse that can be placed in a window for starting seeds or cuttings.

The author made a box and drilled holes in the ends to hold the wire.

EXPERIMENT 2: Since the year 1909, when Dr. Robert Wood published an article in the *Philosophical Magazine,* many scientists have declared the "greenhouse effect" is false. They claim the heat build-up comes from the confining of the air, not on trapped heat vibrations. See if an experiment can be devised to prove whether this is correct.

An Indoor Greenhouse

A CRICKET THERMOMETER

NEEDED: A chirping cricket.

EXPERIMENT 1: Count the number of times your cricket chirps in 15 seconds, add 40, and you have the approximate temperature in degrees Fahrenheit.

REASON: Activities in many animals vary as temperatures vary. Snakes are a good example. In cold weather they are sluggish; in warm weather they slither along at high speed.

The cricket's voice is pleasant to most people, and the little fellow usually does no harm. If he finds a place of some seclusion near a fireplace, he may live there all winter.

EXPERIMENT 2: Crickets may be kept as pets. Put grass sod in a box, cover the box with mosquito netting, and insert the crickets. Keep the grass moist, and feed the crickets bread scraps and fruit rinds.

HOT GRASS

NEEDED: A bushel or more of fresh green grass clippings, crumpled newspapers, two thermometers.

EXPERIMENT 1: Make a pile or basket of crumpled paper the same size as the grass pile. Insert thermometers into each and

A Cricket Clock Thermometer

keep track of the temperatures. At first they are likely to be the same, but by the next day the temperature of the grass may be 20 to 25 degrees higher, day and night, then that of the paper.

EXPERIMENT 2: If there is a cow or horse barn nearby, look at a pile of manure on a cold day. Vapor may be seen rising from it.

Hot Grass

The bacteria in it produce the heat necessary to drive water out in the form of vapor.

REASON: Bacteria begin to work at once in the clipped grass. They are alive, and produce heat as they eat, grow, and multiply. Sometimes a compost pile will get hot to the hand. It can catch fire from the heat it produces.

Fresh damp hay stored in a barn can catch fire and burn the hay and barn.

THE WORLD'S LARGEST MANUFACTURER

NEEDED: House plants.

EXPERIMENT: Water and give them plant food. Watch them grow!

COMMENT: The world's largest manufacturing process takes place in the cells of green plants. The process is called photosynthesis, and is necessary for the continuance of life on the earth.

World's Largest Manufacturer

With light as the energy source and catalyst, the plant takes up water and minerals from the soil, takes carbon dioxide from the air, and makes the food for the world. It is said that leaves use up 100 billion tons of carbon per year, transforming it into life-supporting forms.

According to the *New York Times* (11-13-66) all the blast furnaces in the world make only a half-billion tons of steel in the same period of time.

The first step in food synthesis may be summarized as: carbon dioxide plus water, with the aid of chlorophyll, yields a form of sugar, $C_6H_{12}O_6$ plus oxygen. Minerals are not involved in this step.

FREEZING OF TISSUE

NEEDED: Vegetables such as lettuce, cabbage, celery, carrots.

EXPERIMENT 1: Place the vegetables in a freezer and look at them every few minutes. Some will freeze more quickly than others.

EXPERIMENT 2: To show this with water, place a little water in two drinking glasses. Put a little salt in one, then set both in the freezer. The water without the salt will freeze first.

REASON: Lettuce froze quickest for the author, perhaps because the water in it contained a smaller concentration of chemicals and because it offered more surface cells from which to lose heat. Other vegetables froze at different times; the less pure the water the lower their freezing points.

Freezing of Tissue

A Pinch for Pain

A PINCH FOR PAIN

NEEDED: Someone with leg cramps or other leg pain.

EXPERIMENT 1: Have the person pinch the upper lip as shown in the drawing, and see if the pain vanishes.

EXPERIMENT 2: To suppress a sneeze, press the finger firmly along the upper lip. The sneeze coming on probably will be stopped before it develops. Try it!

REASON: Milton F. Allen, businessman of P.O. Box 789, Decatur, Georgia, suffered severe leg cramps, and discovered that such a pain of the face above the lip relieved the pain. He calls it "acupinch."

Mr. Allen's discovery has been publicized, and he has received hundreds of letters from people who found acupinch a successful treatment of their cases.

Mr. Allen, a religious man rather than a scientist, credits his discovery to prayer he offered when the pain was almost unbearable. He offers no explanation as to why a pinch of the face cures pain in a leg. He says it must be akin to acupuncture. This may not be profound science, but is harmless and interesting to try.

MINERALS FROM THE EARTH

NEEDED: Flower pot, small stones, soil, water, a vessel to catch the water.

Minerals from the Earth

EXPERIMENT 1: Place some stones in the bottom of the pot, and soil above them. Place the pot into or above the container, and pour water into the pot. As it dribbles through the soil it takes up many nutrients that will be helpful in making plants grow.

EXPERIMENT 2: Try using three plants. Put one in garden soil, one in sand, and another in garden soil. Water the first with your enriched soil water. Water the others with distilled water. Compare their growth.

REASON: Plant nutrients must be water soluble, that is, capable of being dissolved in water. Many of them dissolve in the water as it seeps through the soil in the pot.

Try watering house plants with this "enriched" water. They should grow better than plants watered with regular tap water.

NEW PLANTS FROM LEAVES

NEEDED: African violet plant, potting soil, Rootone, water, a clay pot.

EXPERIMENT: Break a mature leaf from the plant, dip the broken end in Rootone, and place the stem in water or in soil. In two or four weeks a new plant will be growing from the broken stem. Cut or break off the new plant—and the same leaf and stem may be used again to start a new plant.

COMMENT: If the leaf is placed in soil put a plastic bag over the pot to hold moisture. Keep away from direct sunlight. If the city water contains chlorine (and it probably does) it is well to let water set overnight in a plastic or enamel container, then heat it slightly and let it cool before using it to water a plant. This will take most of the chlorine out.

New Plants from Leaves

SEE THE PULSE BEAT

NEEDED: A kitchen match, a dime, a candle, a table.

EXPERIMENT 1: Drop wax from the candle on the dime and stand the match in it so the match stands upright. Place the dime on the wrist, and hold the arm and hand still on the table. If the dime is on the right place the match will be seen to move back and forth slightly as the pulse beats.

EXPERIMENT 2: Look for parts of the body where the motion of blood vessels indicate heartbeats.

EXPERIMENT 3: Have a friend exercise, then notice how much stronger the beats show.

REASON: The heart pumps blood through the body in an uneven manner. This can be observed better in some parts of the body than in others. The wrist is a good place to notice the heartthrobs because the blood vessels are near the surface of the skin.

See the Pulse Beat

ALL NATURAL DYES

NEEDED: A beet, a carrot, a squash, beans, cabbage.

EXPERIMENT 1: Boil the vegetables separately. The water in which the beets are boiled will be red; water in which the other vegetables are boiled will remain almost clear.

EXPERIMENT 2: Natural dyes were once used for dyeing all fabrics for our ancestors' clothing. Walnut hulls would give brown dye; brown onion skin, light brown; hickory bark or chips, yellow; green grass, green; red clay, various shades of red. Boil these substances in water to obtain dyes from them.

REASON: Natural dyes give vegetables their color, and these dyes are different in several ways. The natural dye in beets, for example, is readily soluble in water while the dyes in the other vegetables are not readily soluble in hot water.

THE ABSORBING SAND

NEEDED: Three cans, one filled with red clay, another with garden topsoil, another filled with sand, water.

EXPERIMENT: Pack the sand and soil down and pour an equal amount of water on top of each. Watch what happens.

OBSERVATION: The water poured into the sand will pass through it rapidly. Water poured into the clay will stand on top of it a

All Natural Dyes

long time. Water poured into the topsoil will seep into it slowly, and be held there.

REASON: When the earth is stripped down to the clay, the rain that falls on it will run off very fast, and possibly produce flooding. Sandy soil will let the water pass down to harder and more compact layers, possibly too deep to support plant life well. Topsoil holds rain, making it available to plants for good growth.

Once good topsoil is destroyed it cannot be rebuilt easily and quickly. This is good for ecology-minded people to remember.

The Absorbing Sand

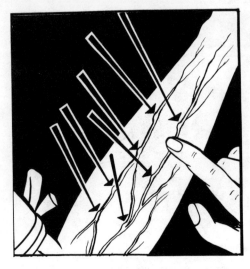

Valves in the Arm

VALVES IN THE ARM

NEEDED: Someone with prominent veins in the arm.

EXPERIMENT: Rub a vein backward, from the heart, and watch little knots form at various places on the vein. These are valves which let the blood flow only in the right direction, through the veins to the heart. Sometimes a tourniquet will make the valves show up. Squeezing the fist helps, too.

COMMENT: Back in 1628 William Harvey used these little demonstrations in proving that the blood circulates. He published his findings in the book *On the Motion of the Heart and Blood in Animals* which may be seen in a library. He upset several of the superstitions about the body, superstitions that were accepted as fact at the time.

TASTELESS COFFEE

NEEDED: Some dry coffee.

EXPERIMENT: Hold your nose and chew some dry coffee—or let some instant coffee dissolve on the tongue. The good coffee taste is not there. Take a breath, and the taste is there.

REASON: The sense of taste is limited to a few basic tastes, and unless a substance affects these there is no taste apparent. But the sense of smell is one of our wonderfully all-embracing senses. It can detect hundreds of different odors.

So closely related are these senses that it is difficult sometimes to determine which we are experiencing. The "taste" of many foods is not taste at all, but odor. The coffee is an example.

Tasteless Coffee

This explains why foods do not taste right when we have a cold. The odors cannot get through clogged sinuses to the sensors of smell.

EGG CANDLING

NEEDED: A cardboard tube or box, an electric lamp, a dark room.

EXPERIMENT: Cut a hole just smaller than the size of an egg in the cardboard, and mount the light some way behind the hole. Hold an egg to the hole, broad end upward, and look carefully.

COMMENT: If the egg contents do not fill the shell the egg is not perfectly fresh. The larger the air space the older the egg. The yolk should be perfectly clear and round in outline. If, besides the air space, there is a dark haze or cloud in the egg, it has spoiled. Any egg from a store will show air space; this does not mean it is not perfectly good for eating. (Suggested by *Organic Gardening and Farming*, April, 1974.)

LAYERING

NEEDED: Rose bush, ivy, or other plants.

EXPERIMENT: Bend the plant down, cover a section with moist earth, and that section will take root and form a new plant.

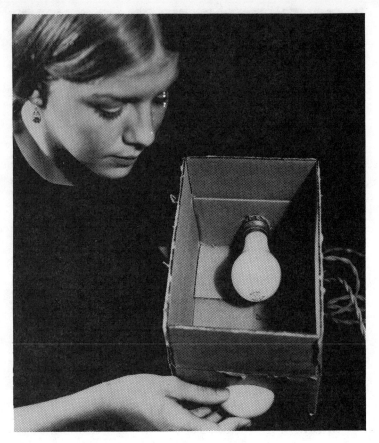

Egg Candling

Layering

COMMENT: Many plants may be propagated in this way. The process is called "layering." Many, such as strawberries, ivy, and Virginia creeper, gooseberries, blackberries, forsythia, carry out the layering process without the help of the gardener.

A grape vine may be bent down, run along in a trench in the ground, and new shoots will rise. Covered with damp soil, the shoots will form roots, and the shoots with roots may be cut away. They are new plants, and the parent vine is not injured.

THINK WARM

NEEDED: A small thermometer.

EXPERIMENT: When the hands are cold, hold the thermometer covered by the hands and think "My hands are getting warmer." See if they get warmer. If the hands are warm, see if, by repeating the suggestion that they are getting cooler, they do get cooler. Repeat the suggestions over and over.

COMMENT: Dr. Edward Taub, psychologist at the Institute for Behavioral Research, Silver Spring, Maryland, taught 19 out of 20 subjects to change their skin temperature by using what he called "thermal imagery" and biofeedback.

Without technical aids, 10-year-old Beth Klinstiver, a fifth grade pupil in Eaton's School near Lenoir City, Tennessee, was able

Think Warm

to change the temperature of her hands three degrees by "thinking warm" alone. She first held the thermometer tightly in her hands 15 minutes to bring the reading to the normal temperature of her hands.

This is a safe psychological experiment for boys and girls to try. (Suggested by an article in *Science News,* October 26, 1974.)

A Cucumber in a Bottle

A CUCUMBER IN A BOTTLE

NEEDED: A growing plant, a bottle, patience.

EXPERIMENT 1: Put the cucumber plant into the bottle while it is still small enough to go in. Let it grow inside the bottle. It will be a curiosity to those who may not know how it was done.

EXPERIMENT 2: Try this with a squash, pumpkin, or gourd plant. Either of these should fill a larger bottle or jug. If one dies after being inserted into the bottle, try another.

COMMENT: If an entire plant is put into a small-mouth bottle it probably will die. The growing cucumber does not have to be in the open; its food is supplied through its stem from the plant which is growing normally in earth and air.

(Suggested by Bob Sherman in *Nature and Science.*)

GREEN LEAVES IN WINTER

NEEDED: Observation.

EXPERIMENT: Watch someone's lettuce rows or grow some lettuce in a pot out of doors.

COMMENT: Some leaves die and fall at the first sign of cold weather. Others freeze as the temperature drops. Some stay green all winter.

REASON: Just why certain varieties of plants tolerate a loss of heat is not always known. In general, an increase in concentration of cell sap lowers the freezing point of the cells. Some plants have less water (cell sap) than others, and when freezing does take place sharp ice crystals do not damage cell walls as much as in other species. Heat of oxidation is liberated more rapidly in some plants than in others at low temperatures and this offsets winter temperatures to some degree. But if temperature drops enough, nearly all growing things exposed to it will die.

The photo shows children in the author's vegetable garden, where all plants had been killed by cold weather except lettuce.

Green Leaves in Winter

Falling Rain or Snow

FALLING RAIN OR SNOW

NEEDED: Observation.

EXPERIMENT 1: While driving along in gently falling rain or snow, notice that the drops or flakes seem to be coming toward you. When the car is stopped they will be seen to fall straight down.

REASON: The appearance of the drops or flakes coming toward you is due to the fact that you are moving toward them, and the motion is relative. You are moving with the car, and while the drops or flakes may be falling straight down, you are getting closer to them as they fall.

EXPERIMENT 2: Often the expression "the rain came down in sheets" is heard. This may be observed sometimes when there is strong wind and a heavy downpour at the same time. The drops are close together as they fall. The wind breaks them into smaller droplets, and together they look almost like solid sheets of water. Watch for this!

EXPERIMENT 3: Drive down a straight road across which snow is drifting. A strong wind seems to be blowing the car in the direction in which the snow is drifting, and you feel that your muscles must fight against a real force.

FALLING LEAVES

NEEDED: Observation.

OBSERVATION: A thin stem called the petiole connects the leaf to the main stem of the plant. Where the connection is made there are special different cells called abscission cells. While the leaf is growing and healthy a chemical produced in the leaf, indoleacetic acid or auxin prevents the abscission cells from forming into a cork-like mass. When the summer is over, this mass forms, separating the growing parts of the plant and letting the leaves fall. A scar may be seen where a petiole breaks off, and this scar is made up of abscission cells.

Falling Leaves

BALANCING

EXPERIMENT 1: Stand on one foot. It is easy. Stand on both feet, close the eyes, then lift one foot. It is almost impossible for most people to stand on one foot with the eyes closed, for more than a few seconds.

EXPERIMENT 2: Try running with the eyes closed. It may be difficult. Be prepared to fall down! A soft grassy place is best for trying this.

REASON: We have balancing organs in our ears. The three semicircular canals are looped tubes placed at right angles to each

Balancing

other and filled with liquid. The tubes are lined with nerve cells connected to the brain. Also, tiny loose particles touch nerve endings when the head is tilted.

But in addition to these specialized organs we depend on other senses to keep our balance. Our eyes are important for this; also, our muscles detect any swaying movement and send messages to the brain to correct it.

Keeping our balance with eyes closed is easier after practice. Also, some people find it easier than others.

COMMENT: One theory of seasickness is that the various balancing signals we use, those from the ears, from the muscles of the legs and trunk, from visual clues (we *know* that the corners of a room are vertical) in the unusual experience of being tossed around in a ship, give conflicting and contradictory messages. The unaccustomed effort of having to decide which signals to believe and act on results in tension, which results in sickness, or malaise.

HOLDING THE BREATH

NEEDED: Observation.

EXPERIMENT: Hold the breath as long as possible. It is painful. A time comes, and very soon, when holding the breath is no longer possible.

REASON: The reasons are not fully understood. But two little glands, called the "carotid bodies" are thought to be the parts of the body that react to breath-holding and send out the distress signals.

The carotid bodies are sense organs about five millimeters long in a man. They are located largely behind the carotid artery in the upper part of the neck. They are sensitive to reduced levels of oxygen in the blood. They are responsible for the increased rate of breathing at high altitudes where the air is thin.

A group of scientists in Harbor General Hospital in Torrance, California, is studying the problem, according to the magazine *Scientific American,* June, 1974.

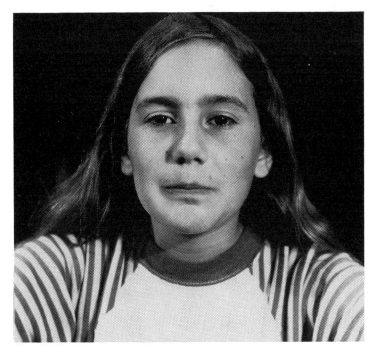

Holding the Breath

SPALLANZANI

NEEDED: Two jars with tight lids, boiling equipment, juices.

EXPERIMENT: Place juice in two jars. Boil one as in canning; leave the other cold. Seal the boiling jar while hot. Set both jars aside; the unboiled juice will sour or mold or both.

COMMENT: Lazzaro Spallanzani, Italian biologist who lived 1729 to 1799, showed in this way that air carries microscopic life and

Spallanzani

that boiling kills the bacteria. That was one of the great discoveries of science.

Spallanzani was the first person to watch bacterial cells divide. He discovered that bats do not need to see in order to dodge wires, although he did not solve the mystery. He discovered that some animals, including salamanders, can grow new legs to replace any that are lost.

He was a teacher at the University of Padua.

REMOVE WARTS

NEEDED: A pencil with rubber eraser.

EXPERIMENT: Tell a little brother or sister "I'll make your wart go away. Each time I touch the wart with the eraser you must say 'Wart, you must leave me.'" Then touch the wart lightly and slowly six times with the rubber. The wart may go away overnight.

REASON: This is a psychosomatic cure that is not understood, but doctors know it does work in many cases. Any objects may be used to so "charm" the wart away; a coin may be rubbed on it, for example. The charm works less frequently on older people who may be very skeptical.

If warts cannot be charmed away, a doctor can touch them with a little mild acid, or "burn" them off with high frequency electricity that does not shock. The doctors' methods always work.

One common folklore wart remedy is to steal a neighbor's dishrag, rub it across the warts, and bury the rag in the woods. Another involves a dead cat in a graveyard at midnight. Another involves stump water, water gathered in a hollow stump. You immerse your hand with the warts and say, "Stump water, stump water, take away warts."

Warts are now known to be caused by (communicable) viruses. Most adults have become immune. Hence, warts are usually a childhood condition.

Nobody knows how belief in a particular applied magical cure can do anything to a virus, but it is well-established that it can.

Remove Warts

GROW A PINE TREE

NEEDED: A pine cone with seeds, a flower pot, and soil.

EXPERIMENT: Plant a seed from the pine cone, twice as deep as the diameter of the seed. Keep the dirt moist until the seed sprouts. Then place it in direct sunlight and water it about once a week.

Grow a Pine Tree

COMMENT: Suzanne Brown, the author's granddaughter, found such a cone, took a seed out with a knife, and planted it. It grew. This is a rather unusual experiment for a boy or girl, growing a tree from a pine seed.

Cone-bearing trees, or "conifers" are interesting to study. An encyclopedia has a good section on them. There are two kinds of cones on the same tree, pollen cones and seed cones. The seeds are not enclosed, but are free to fall out from between the scales of the cones. They fall to the ground and some of them grow into trees.

Cones can be as large as 15 inches, or can be very small, depending on the type of tree or shrub.

HOW FAST ARE YOUR SIGNALS?

NEEDED: A friend.

EXPERIMENT: Have someone inflict slight pain, and see how quickly the signal travels through nerve fibers to the brain.

COMMENT: All impulses travel so quickly that a boy or girl could not determine which kind travels fastest. But Henri Busignies, writing in *Scientific American*, says that each bundle of nerve fibers has its own characteristic conduction speed, from a few meters to about 100 meters per second.

Signals conveying muscle position travel at the highest velocity, he says, presumably because balance and quick movement are vital. Pain signals are among the slowest.

A pinch is a harmless experimental pain.

How Fast Are Your Signals?

STERILIZE SOIL

NEEDED: Soil to be sterilized, baking bags, an oven.

EXPERIMENT: Place the soil in the bags, seal them, and heat in the oven an hour or longer at a 250-degree temperature.

REASON: Soil for potting may contain a harmful fungus, or a parasite, that can attack the plants as they grow. Sterilizing at this temperature kills any living thing in the soil. Of course, the soil must cool before a plant is placed in it.

Sterilize Soil

Nematodes, fungi, and weed seeds are killed by sterilizing. A too-high temperature is not desirable, however, because it can destroy some of the nutrients of the soil.

Watch the Roots

WATCH THE ROOTS

NEEDED: Glass jar, white blotting paper, cotton or cotton rags, seeds, water.

EXPERIMENT: Cut the blotter so it fits around the inside of the jar. Place it in the jar, and hold it in place by stuffing cotton or rags loosely behind it. Use a pencil or knitting needle to open space for placing the seeds inside and arranging them, as shown. Keep the blotter moist.

OBSERVATION: The seeds will sprout and the roots and stems may be seen clearly as they grow. By turning the jar upside down the roots and stems will be seen to reverse their direction of growth.

PLANTS AND TEMPERATURE

NEEDED: Two boxes the same size, topsoil, grass sod, two thermometers.

EXPERIMENT: Place soil in one box with a thermometer, place soil with sod above it in the other box with a thermometer. Place both boxes in the sun. The temperature will be higher in the box without the growing sod.

Plants and Temperature

REASON: The leaves of grass—and other green leaves—keep cool by bringing up water from the roots and having it evaporate from the leaves. The evaporating water cools the air.

This explains partially why wooded land is cooler than open fields or deserts. Another reason is that the shade from trees keeps the sunlight from hitting the ground and turning into heat.

When the author tried this experiment the temperature settled at 89 degrees in the bare soil and 82 degrees in the sod after five hours in the sun.

The bare soil gets colder quicker at night.

Tongue Foolers

TONGUE FOOLERS

NEEDED: Grapefruit juice, water, cocoa or candy, salted peanuts.

EXPERIMENT: Drink a sip of water, then some grapefruit juice. The juice is delicious. Take a sip of cocoa or eat a small piece of candy, then taste the juice. It is very sour. Eat some peanuts or other salty food, and taste the juice again. It is again delicious.

REASON: The juice is acid, and therefore sour, but only mildly sour—and pleasant. Immediately following candy or sugar the same juice seems to be unpleasantly sour by comparison. But the immediately previous sweet taste is offset by more grapefruit juice or salty food.

If you eat artichokes, Cynara scolymus, all things will taste sweet afterward, even water.

There is an African fruit called miracle fruit that has this property in very high degree. If you eat this fruit, afterward even a lemon will taste as sweet as a tangerine, and quite similar.

AFTERIMAGE, AN EYE TRICK

NEEDED: Black paper, white paper, scissors, paste, a strong light.

EXPERIMENT: Cut a disc two or three inches in diameter from the white paper. Cut a square twice as large from the black paper. Paste the white disc on the middle of the black paper.

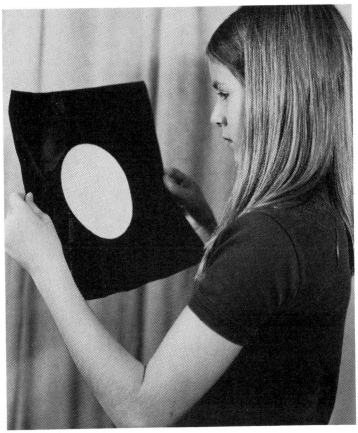

Afterimage, an Eye Trick

Place the papers in strong light and gaze at the white disc for one minute. Then look at a blank wall that is not brightly lighted. The eyes will see the disc again, but this time it will be black.

REASON: When the eyes are exposed to bright light for several seconds the retina shows tiring or fatigue of those retinal cones sensitive to the color observed. When the same area of the retina observes another surface such as the wall dimly lighted, the other retinal cones respond more strongly and the complementary color is seen.

"Afterimages" of this kind cause black to become white, white to become black, yellow to become blue.

DON'T BLINK

NEEDED: Two people.

EXPERIMENT 1: Puff breath at the top of a nose, and dare the person to stare and not blink. It is almost impossible to refrain from blinking when the air hits the eyes.

EXPERIMENT 2: Hold a thin sheet of plastic in both hands, and bring it quickly toward a friend's face. The friend knows it could not do harm, yet he will blink. This is a protective reaction.

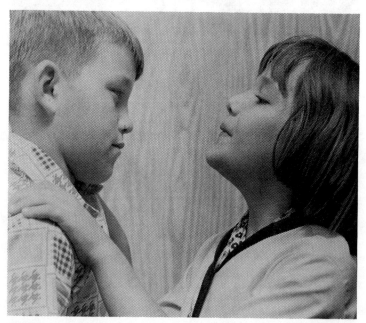

Don't Blink

Eye Games

EYE GAMES

NEEDED: A table, two objects, a friend.

EXPERIMENT 1: Have the friend kneel so his eyes will be on a level with the table top. Place one object in the middle of the table, so it does not make a shadow. Have the friend hold a hand over one eye. Place the other object a few inches closer or farther from the friend, and have him guess which it is.

EXPERIMENT 2: Try this with the objects twice as far away from the eye. Also, try it with small objects closer to the eye.

REASON: Our eyes turn slightly inward when they focus on an object, and the brain interprets the turn to show us the distance of the object. This is called stereoscopic vision.

When only one eye is used there is no stereoscopic vision, and it is difficult to tell distances. In this experiment it will be difficult for the friend to tell whether the second object is closer or farther than the first.

COMMENT: Nearly all predatory mammals have excellent stereoscopic vision, their eyes in the front, looking straight ahead. Most of the day-hunting raptorial birds, such as hawks and eagles, have excellent stereoscopic vision.

Non-predator mammals, such as rabbits and field mice, have their eyes on the sides of the head. A rabbit has almost 360° vision (in all directions). The only blink spot in the rabbit's vision is a small cone containing his ears.

Starchy Crackers

STARCHY CRACKERS

NEEDED: Crackers, household tincture of iodine, water, a dish.

EXPERIMENT 1: Add two drops of the iodine to a half glass of water. Wet one cracker, and place it on the plate. Chew the other cracker until it tastes sweet, and place it from the mouth to the plate. Put a few drops of the water on each cracker.

EXPERIMENT 2: Chew a piece of highly flavored chewing gum, take it out of the mouth, then chew the cracker. Is there a difference?

EXPERIMENT 3: Start chewing the cracker with a mouth full of soft drink. Is there a difference?

OBSERVATION: The iodine solution placed on the wet cracker turns it a purplish color; that placed on the chewed cracker does not change its color.

REASON: The purple color is a test for starch, showing that the cracker contains starch. Saliva has changed the starch in the other cracker to sugar, so that there is no color change. Digestion of starch begins in the mouth. And continues in the stomach.

HYDROTROPISM

NEEDED: A cardboard box, screen wire, soil, lima bean seed, a small dish for water.

EXPERIMENT: Make a hole in the bottom of the box and place the screen wire over it. Put a half inch of dirt in the box, plant one or two seeds in it, and cover with another half inch of dirt. The soil should be kept moist. Place the box over the dry dish or saucer.

The roots will start growing downward in the normal way, responding to the pull of gravity. This is called "geotropism." But when they have grown through the screen they turn to the side, seeking water. This seeking of water by the roots is called "hydrotropism."

REASON: The attraction of roots toward water is stronger than their attraction downward.

The experiment has been performed without water below the box. Now try it again, this time placing a dish of water under the box, so the water comes to within a quarter of an inch of the screen. The roots will grow downward into the water.

Hydrotropism

GEOTROPISM

NEEDED: Milk cartons, dirt, bean or mustard seed.
EXPERIMENT: Cut off the cartons to make flower pots, plant the seeds in them. Let the seeds sprout until the plants are an inch or

Geotropism

two high, then turn some of the cartons on their sides, leaving the others upright.

The earth's gravity causes the roots to grow downward and the stems upward in each case. The root growth may be seen if the cartons are cut away from the dirt.

In the lower drawing, the plant is growing upward from the overturned pot. In the upper drawing the plant has righted itself after the pot was turned upright again.

PLANT FORCE

NEEDED: Pots, dirt, lima bean seeds, six quarters or other weights.

EXPERIMENT: Plant the seeds, one to a pot. Tape two quarters together, and three quarters. Place one quarter over one seed, two taped together over another seed, and three over the other. The force of the growing plants will push all the weights aside.

REASON: The division of cells that we call growth can exert tremenduos force. Growing roots can break rocks apart. This force is developed when cells in growing tissues split and enlarge as water and nutrients are absorbed and used to make new cellular materials. "Osmotic turgescence" is the term frequently used to indicate the chief forces involved. (The pots shown in the drawing are made from the bottoms of milk cartons.)

PHOTOTROPISM

NEEDED: Milk cartons, soil, bean or mustard seeds.

EXPERIMENT: Cut off the cartons, place dirt in them, and plant the seeds. Place some cartons in a dark place, some where bright light reaches them from one side and others where light is about equal on all sides. The plants will grow toward the brighter light.

REASON: Cells on the darker side of the plant grow faster and longer. The seeds planted in a dark place will grow faster, but will die in a few days for lack of food. Growth regulating substances called "auxins" are produced in the tips of young growing stems, and migrate in greater quantity toward the darker side of the stems.

A TRICK OF THE EYE

NEEDED: A card, a pin, an electric bulb.

EXPERIMENT: Make a pin hole in the card and look at the light through it. Bring the pin near the eye and into the line of sight between the hole and the eye. The pin will appear upside down.

REASON: The simplified diagram shows the paths taken by the light rays from the bulb to the eye. The pin in this case is a

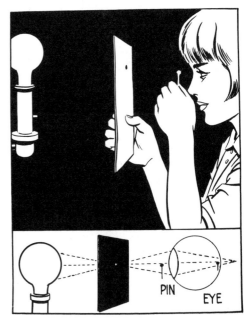

A Trick of the Eye

shadow cast on the retina of the eye. Since the shadow is right side up, the brain will see it as being upside down.

This may not be easy to see on the first try. Hold the pin close to the eye. Use a bright frosted bulb. Hold the card at arm's length.

Another Trick of the Eye

ANOTHER TRICK OF THE EYE

NEEDED: Cardboard and a heavy needle or pin.

EXPERIMENT: Make a hole in the card with the needle, then have someone who reads with glasses remove the glasses, close one eye and read. Then have him hold the card close to the eye, looking through the hole to read. The printing will be much clearer.

REASON: Because of eye defects the light rays entering the eye, and shown in dotted lines, do not focus on the back of the eye, the retina. The eyeglass lens corrects the fault so the rays all focus correctly. But if the only light entering the eye passes through the center of the cornea or lens it does not have to be focused. It goes straight through to the retina and produces the image, as shown in the heavy line.

THE GREAT REDI EXPERIMENT

NEEDED: Three wide mouth jars, three pieces of raw meat, paper, cheesecloth.

EXPERIMENT Place a piece of meat in each jar; cover one with paper, cover one with two thicknesses of cheesecloth, and leave the other uncovered. The meat in all three will putrefy, but maggots will appear only in the uncovered jar. (Leave the jars out of doors in warm weather for this experiment.)

REASON: Aristotle taught that maggots appeared spontaneously in putrefying flesh or filth. Francisco Redi, in 1668, performed this experiment (in a much more elaborate fashion) to prove that life can come only from other life; in this case, maggots can come only from fly eggs.

It may be necessary to protect the jars from night-prowling cats. The author used screen wire with large mesh for this. The mesh must be large enough to allow flies to get through it.

THE PRESSURE OF SWELLING SEEDS

NEEDED: Lima bean seeds, dry sand, water, a pint jar.

EXPERIMENT: Mix half beans and half sand in the jar, shaking the jar and pushing the sand in tightly. Wet the sand, but do not put in enough water to flood it. Screw the lid on tightly (it does not have to be air tight, however).

The beans will absorb water from the spaces between the sand grains, swell and in a few hours burst the jar. Be sure the jar is placed on a large cookie sheet or other sheet to catch the broken glass and sand.

AN UNUSUAL TOUCH SENSATION

NEEDED: Two people.

EXPERIMENT: Hold the fingers as shown and rub them with the finger and thumb. The feeling will be a totally unfamiliar one.

REASON: For years we have grown accustomed to touching our bodies in various ways and getting the same sensations each time. If the other person's finger is inserted in a place where we are accustomed to feeling our own finger the difference in the pattern of feeling registered in our brain is quite astonishing. We are all creatures of habit, and one of our important habits is our consistent and dependable sense of touch. In this experiment we are playing a trick on our sense of touch.

SEE ROOT HAIRS

NEEDED: A saucer, a dish, paper towel, radish seeds, water.

EXPERIMENT: Cut a piece of paper towel the diameter of the saucer, put it in the dish, and wet it. Put radish seeds on the wet

An Unusual Touch Sensation

paper, half an inch apart, invert the saucer over the paper, and leave it for two days in a warm place. The white cotton-like growth seen is made up of many root hairs.

A root hair is a long narrow outward extension of a single surface cell. The root hairs extend into the soil between earth particles, to take up moisture and nutrients for the growth of the plant. A magnifying glass will show them more clearly.

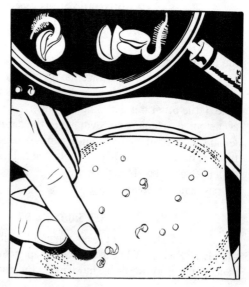

See Root Hairs

INFECTION

NEEDED: Two good apples, one apple with a rotten spot, needles, matches, string.

EXPERIMENT: Tie the string around the stem of one healthy apple to identify it. Sterilize both needles by heating them in the flame of a match. Let them cool. Stick one needle into the rotten spot, then without cleaning it, stick it into the apple with the string marker. Stick the other needle into the other sound apple, without first sticking it into the rotten spot. Remove both needles, throw away the rotten apple, and place the other in a warm room.

If other factors are equal, the apple with the string should develop a rotten spot where the needle penetrated it.

REASON: Bacteria will have been transferred from the rotting apple to the good one.

PUFFBALLS

NEEDED: A puffball, found in woods or on the lawn.

EXPERIMENT: Kick the puffball lightly with the foot. "Smoke" puffs up from it.

REASON: That "smoke" consists of milllions of spores of that particular mushroom plant. They are carried by the breeze, fall to the ground, and are likely to produce other plants in other locations.

Puffballs

Some puffballs may be three feet in diameter. Mushrooms (and puffballs) are fungi, plants that do not produce their food but live on food taken from decaying matter or sometimes from live hosts. "Devil's Snuffbox" is the name sometimes given to the puffball.

A TEST FOR VITAMIN C

NEEDED: Half a teaspoonful of cornstarch, water, tincture of iodine, foods to be tested (perhaps lemon and orange juice).

EXPERIMENT: Boil the starch in half a glass of water. Put 20 drops of the mixture into a glass of water, and add one or two drops of iodine. A blue color should appear. If food containing vitamin C is added (drop by drop), the blue color should disappear.

REASON: Food containing vitamin C seems to destroy the somewhat mysterious combination of starch and iodine which is responsible for the blue color. But the color is also destroyed in other ways, so this is not a specific and sure test for vitamin C, in spite of some claims.

Yeast and Carbon Dioxide

YEAST AND CARBON DIOXIDE

NEEDED: Two large jars, small glass, molasses, yeast, lime.

EXPERIMENT: Put limewater into the glass and set it in the large jar. Put a small amount of molasses and yeast in the jar, and put in water until it comes almost to the top of the glass. Screw the lid on the jar, let it sit in a warm dark place, and in two days the limewater should be milky rather than clear—a test for carbon dioxide.

REASON: Yeast cells contain enzymes which change the sugar of the molasses to form grain alcohol and carbon dioxide. The carbon dioxide unites with the limewater, producing a small amount of calcium carbonate which stays in suspension, forming the white or milky color. Chemically the limewater test is:

$Ca(OH)+CO_2 \quad CaCO_3+H_2O$.

Limewater may be bought in the drug store or made by mixing lime and water, letting the mixture settle until clear liquid may be poured off the top. As a double-check or "control" put a glass of limewater into an empty closed jar at the same time. The limewater will remain clear.

This limewater test can be tricky. If too much carbon dioxide goes into the solution the calcium carbonate becomes calcium bicarbonate, which may dissolve and leave the solution clear again.

MAKE A KILLER JAR

NEEDED: A jar with a lid, a small piece of screen wire, cotton, some nail polish remover, a small insect.

EXPERIMENT: Place the cotton in the bottom of the jar. Cut the screen and bend it so it will fit down around the inside of the jar, over the bottom. Soak the cotton with nail polish remover and drop

Make a Killer Jar

the insect into the jar. Close the lid tightly. If the bug does not die in a few minutes, add some more remover. This way of killing insects lets them die relaxed, and their appearance is unharmed.

REASON: All animals and insects require oxygen. Here the oxygen supply is replaced in the air with polish remover fumes which are poisonous in dense concentrations.

EARTHWORMS

NEEDED: A jar, moist earth, sand, cornmeal, dark paper or cardboard.

EXPERIMENT: Dig some moist earth from a place where earthworms are found. Fill the jar about two-thirds full of soil, sprinkle some cornmeal on it, then cover with half an inch of moist sand. Wrap the paper around the jar so no light enters. Keep the top open so the earthworms can get air.

After several days the burrowing worms will have brought earth to the top of the sand. Some of their burrows may be seen against the inside of the glass. Earthworms are harmless and very beneficial in the soil. They continually mix and enrich it, and loosen it so that air may circulate through it.

THE POWER OF SUGGESTION

NEEDED: Two people.

EXPERIMENT: Do not let the other person know that an experiment is being conducted. Scratch your head with a finger, and

Capture Spider Webs

chances are the other person will do the same. Cough, and the other person is likely to clear his throat or cough.

REASON: The power of suggestion is very strong. We are all subject to it. When the other person sees us scratch, he immediately thinks he needs to do the same. When we cough, we merely suggest it to the other person, and he subconciously acts on the suggestion.

CAPTURE SPIDER WEBS

NEEDED: A duster, black paper, sifted flour.

EXPERIMENT: Place flour in the duster. Find the spider web in early morning when dew is on it. Dust some flour on it so all the strands are covered lightly. Bring the black paper up behind the web, and lift the web gently on the paper. It will stick.

To make the duster punch a dozen holes in the bottom of an old plastic bottle. The flour is dusted out as the bottle is squeezed.

Chapter 6
Water & Surface Tension

HOLES IN THE WATER

NEEDED: A pan of water, four identical blocks of wood, wax paper.

EXPERIMENT: Wrap two of the blocks with wax paper (thumb tacks may be used to hold the paper around the blocks). Place the two unwrapped blocks close together on the water, and they will come together. Place the wrapped blocks on the water. They do not come together; they may push farther apart.

REASON: Notice that the water wets and clings to the unwrapped blocks as in the diagram, actually pulling itself up along the sides of the blocks. The surface tension of the water between the blocks, acting like a stretched rubber sheet, pulls the blocks together.

The wax paper is not "wetted" by the water; it seems to cause the weight of the blocks to actually make holes in the water. The lower diagram shows how this looks on close observation.

STRETCH A WATER SURFACE

NEEDED: A glass of water and a small stiff wire.

EXPERIMENT: Place the wire under the water surface and bring it up slowly as in the large drawing. If care is taken, and the wire moved very slowly, the surface of the water may be seen to stretch upward before the wire finally breaks through it.

REASON: The surface tension of water is much like a stretched rubber sheet. It actually takes a little force to break

Holes in the Water

through it, just as it would take a greater force to pierce a rubber sheet.

If the experiment does not work, try greasing the end of the wire with butter or oil.

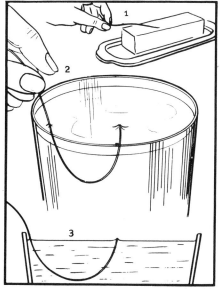

Stretch a Water Surface

SOAPY SMEAR

NEEDED: Pieces of flat glass, soapy water, plain water, two atomizers.

EXPERIMENT: Spray the soapy water on one piece of glass, the plain water on the other. The plain water will form tiny drops; the soapy water will run down more readily and drip off the glass.

REASON: Surface tension tends to make a liquid assume the shape which has the smallest surface area, which is the sphere. This is seen when the plain water is sprayed onto the glass.

Soap or detergent lowers the surface tension, and gravity pulls the liquid downward as a thin film.

WETTER WATER

NEEDED: Water, talcum powder, liquid detergent, a needle.

EXPERIMENT 1: Lower the needle gently to the surface of the water. It can be made to float there. Pour a little detergent on the water, and as it reaches the needle, the needle sinks.

EXPERIMENT 2: Clean the container well, put water into it again, and sprinkle talcum powder on the surface. Pour on some detergent, and it will make a path through the powder as shown.

EXPERIMENT 3: Clean the container again. Put powder and the needle both on the water surface. Watch them sink as detergent reaches them together.

Wetter Water

REASON: The surface tension, a surface film on the water, is strong enough to support the needle. The detergent breaks the film.

More Experiments: Use pepper instead of talcum powder. Use soap instead of regular detergent. (Soap is a detergent, too, but is weaker than regular detergents from the store.)

OILY SMEAR

NEEDED: Window glass, oil, water, a spatula.

EXPERIMENT: Place only a touch of an oil drop on the glass, also several drops of water. Smear with the flat edge of the spatula, and the fogginess of the windshield wiper in a light rain can be duplicated.

REASON: On the windshield, the small amount of oil usually comes from drops of water tossed up from the highway by tires of other vehicles—rarely from the wiping cloths at filling stations.

Water on an oily surface forms into tiny round balls if there is a very small amount of it. This is because surface tension tends to make the water droplets take the shape of spheres. When the rubber spatula or blade is rubbed over the water it simply divides the drops into many smaller droplets.

The fogginess clears as the blade moves along, because the amount of water involved is so small that it evaporates quickly.

THE SWEATING JARS

NEEDED: Two jars, ice water, hot water.

EXPERIMENT 1: Put hot water into a jar, filling it half full. Note that droplets of water appear on the inside of the glass above the water.

EXPERIMENT 2: Put ice water into a jar, filling it half full. Note that droplets of water form on the lower half of the jar, where the ice water cools it below room temperature.

REASON: Vapor from the hot water rises in the jar, and as it comes into contact with the cooler parts of the glass some of it condenses out and sticks to the glass in droplets.

Air can hold more moisture when warm than when it is cold. As warm air from the room touches the cold glass and becomes cooler, it cannot hold as much moisture and must give up some. Part of what it gives up forms droplets on the glass.

The same principle applies to both of these experiments.

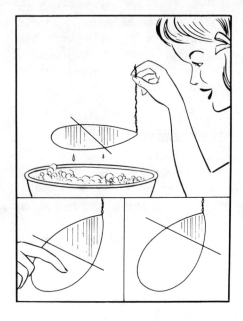

The Mysterious Needle

THE MYSTERIOUS NEEDLE

NEEDED: Wire, pliers, some strong suds.

EXPERIMENT: Make a wire loop with a handle, and cut another piece of wire to rest across it. The piece should be very straight, and all the wires should be clean and smooth.

Place the short piece on the loop, dip both into the suds, and lift them out. There should be a water film covering the entire loop. Break the film on one side of the wire with the finger, and the remaining film will draw the wire toward the edge of the loop.

REASON: The surface of ordinary or soapy water acts as a stretched rubber sheet. This is surface tension, and it is strong enough to pull the small wire or needle along.

EROSION OF SOIL

NEEDED: Soil on a board, an equal area of grass sod on a board, a water hose.

EXPERIMENT 1: Spray water on the soil, and it washes into gullies and may all wash away. Spray the sod the same way, and it resists the washing action of the water.

EXPERIMENT 2: Notice that farmers make use of "contour plowing" to prevent washing of the soil. This means that plowing follows the contours of the land, never allowing rain to wash directly down hill in a plowed furrow.

An experiment can be made to show this if a large square of sod is available. Arrange it so that low places follow a contour, while in another section allow the low "furrows" to run up and down the hill. Water will be seen to flow down much more slowly on the contoured sod.

REASON: The grass softens the blows hit by the drops of water, and the tangle of roots softens the flow of water through the soil underneath. This shows why soil stripped of its grasses or trees soon washes away.

Erosion of Soil

WACKY WICK

NEEDED: Two jars, one filled with water, a cloth, a wire.

EXPERIMENT 1: Roll the cloth around the wire. The wire is to give it stiffness. Bend it. The water will move through the cloth from the upper to the lower jar.

EXPERIMENT 2: This syphon can be used as a filter. Put slightly muddy water in the upper jar, and as it syphons over into the lower jar most of the mud is left behind in the cloth. This is because the particles of mud do not flow through the cloth as does the clear water.

While this is a filter of sorts, it does not purify the water enough for drinking. Look, but don't drink.

REASON: Capillary action, in which the water molecules cling to the tiny fibers of the cloth, causes the water to rise into the cloth against the force of gravity. When the water gets over the edge of the glass, however, both gravity and capillarity combine to pull it downward into the second glass.

"Capillary" comes from a Latin word which means hair, and refers to the tiny bores in "capillary tubes." These are tubes in which water will rise against the force of gravity, the height of the rise depending on the smallness of the bore.

MELTING UNDER PRESSURE

NEEDED: Two ice cubes

EXPERIMENT: Squeeze the ice cubes together. Some of the ice will melt under the pressure, then should freeze again when the pressure is removed, sticking the cubes together.

REASON: Notice that *dry* ice and snow are not too slippery. Actually, the skater skates on water, because the pressure of the steel blade on the ice melts a little of the ice, to provide the slippery water layer. Of course, the water ordinarily freezes again when the skater has gone and the pressure is removed, if air temperature is below the freezing point of water.

MARRIAGE OF THE WATER DROPS

NEEDED: A piece of clean glass and some water drops.

EXPERIMENT: Place drops close together on the glass. They will move closer until they join together.

REASON: Molecules of water attract one another. Gravity flattens the drops on the glass until they touch each other; surface tension then pulls them to a common center of mass. The surface tension, which acts something like a stretched rubber sheet, combines the separate drops so that they have a shape with the least possible surface.

The surface tension is greater than the force of cohesion between the water and the glass.

Dirty glass would reduce the surface tension of the water.

AN OIL DROP ENGINE

NEEDED: A pan of still water, a piece of cardboard cut as shown, a drop of oil.

EXPERIMENT: Place the boat on the water. Place a drop of oil carefully in the hole. The boat will move forward.

An Oil Drop Engine

REASON: The oil, which is lighter than the water, floats on the surface. As it runs out the rear opening, the surface tension is reduced and the surface tension at the front is not reduced. The boat moves in the direction of the greater pull.

Could this be an example of "action-reaction" also? If the oil flows out the back, reaction would tend to propel the boat forward. The rule is that for every action there is an equal and opposite reaction.

WHY THE ROUND DROPS?

NEEDED: A glass of water, rubbing alcohol, castor oil, a medicine dropper.

EXPERIMENT 1: Mix water and alcohol until it is the same weight as the oil. A little experimentation will show when the right proportions of water and alcohol are reached. If the drops rise, add more alcohol. If they move downward, add more water. Drops of oil then, when placed under the surface, remain in their places and assume a round shape.

EXPERIMENT 2: Use a clean glass or jar. Place water in it, and let the alcohol down slowly over the water so they do not mix very much. The oil may then be placed between water and alcohol, and will float there.

REASON: The surface tension of the oil pulls the drops into the shape which has the smallest surface area, which is the sphere.

Why the Round Drops?

MERGING STREAMS

NEEDED: A tin can, a nail, hammer, water.

EXPERIMENT: Make three holes in the can near the bottom. They should be about 3/16 of an inch apart. Fill the can with water, and three streams will come out. If the streams are pinched together they merge into one. Cover the middle hole momentarily with the finger, and three streams will form again.

REASON: The molecules of water cling together—cohesion. When the streams are united, surface tension also comes into the picture. This is like a film stretched over a water surface, and the streams do not break through it easily.

THE LIVELY SOAP

NEEDED: A dish of clean water and a tiny piece of soap.

EXPERIMENT: Use a piece of soap half the size of a pin head. Drop it on the water, watch carefully, and the soap will move around quite fast.

REASON: The surface tension of the water acts like a stretched rubber sheet. The soap reduces the tension, but is not likely to do so evenly. The tension pulls the piece along on the side where it has acted least to affect the surface tension. The movement stops when there is an equal amount of soap over all the surface.

A CLOTH AND SPONGE MYSTERY

NEEDED: A pan of water, a sponge, a strip of cotton cloth.

EXPERIMENT 1: Place the sponge in the water, with the cloth folded on top of it. The water will rise rather quickly to the top of the sponge, but much more slowly through the cloth.

EXPERIMENT 2: Try placing another sponge above the first in place of the cloth. Will the water rise into the second sponge more quickly than it did through the cloth?

EXPERIMENT 3: Try using dry sponges, placing one in the water and the other above it, but this time place a paperweight or other small weight on the second sponge. This should put the sponges into closer contact with each other, allowing the water to flow more easily into the top sponge.

REASON: In the sponge, the fibers are close together, so that the water can rise easily through the sponge by the process known as capillary action. There is no close network of fibers linking the folds of cloth, and so the water practically stops there. Capillary action allows oil to rise in the wick of a lamp, or water to move up through the ground.

A Cloth and Sponge Mystery

WHAT HOLDS UP THE CLOUDS?

NEEDED: A milk bottle, some hot water, an ice cube.

EXPERIMENT: Put hot water into the bottle, place the ice cube over the mouth, and small clouds will be seen to form inside the bottle. What holds them up?

Actually, if left still and not disturbed by air currents, all clouds would eventually fall in air at a constant temperature. But the droplets of water in clouds are so small and light that they do not fall rapidly because of the air resistance. The slightest air current can blow them along. If they join together and become heavier they fall faster, usually as rain.

Rising currents of warm air give an upward push to the tiny droplets and the clouds move up or down according to the amount of the upward-moving air. Watch fog carefully, and you can often see it fall.

HOW PURE IS THE WATER?

NEEDED: A clean glass filled with water.

EXPERIMENT: Let the glass stand on a shelf until the water has evaporated. Notice that the glass will not be clean.

REASON: Ordinary water contains many substances in a dissolved or suspended state. Most of them do not evaporate, so, as the water evaporates or goes into the air as vapor, it leaves the other substances behind. They cling to the glass sides and bottom.

This is not an accurate check on impurities in the water, since some of the residue seen as the water leaves the glass consists of substances that have settled in the water from the air as the glass stood on the shelf.

Such "impurities" as we find by this method do not adversely affect the water as we drink it or otherwise use it. Many of them are good for health. Special tests are needed if water is thought to be unsafe and these tests are not performed at home.

A WATER AND WEIGHT MYSTERY

NEEDED: A container of water, a scale for weighing, a block of wood.

EXPERIMENT 1: Place the empty container on the scale, weigh it, then put the block of wood into it. The additional weight registered is that of the wood. Take the block out.

EXPERIMENT 2: Put water in the container, note the weight, and again add the wood. Again the added weight is that of the wood.

EXPERIMENT 3: Press the wood down with the finger tips until it is barely under the water. This time the added weight is more than that of the wood. Why?

REASON: Archimedes discovered more than 2,000 years ago that an object immersed in water but not touching bottom adds

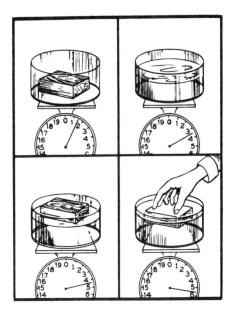

A Water and Weight Mystery

weight equal to that of the water displaced. When the wood is pushed down by force from the finger tips until it is under water more water is displaced.

SHIMMERING LIQUIDS

NEEDED: Water in a flat pan, some rubbing alcohol.

EXPERIMENT: Pour a little of the alcohol into the center of the pan of water. A beautiful shimmering effect is seen moving outward toward the edges of the pan. The shimmering continues until the liquids are mixed.

REASON: Each liquid when separate has its different surface tension. As the alcohol spreads outward, the surface tension of the alcohol-water surface becomes less than the surface tension of the water, so that there is an unbalanced pull. This causes the slight outward shimmering motion which continues until the surface tension pulls are equalized.

WATER ON THE WALL

EXPERIMENT: Notice how water drops form on the wall of the bathroom when you take a shower.

REASON: Warm air will hold more water than colder air. The warm water in the shower warms the air and at the same time offers an ideal condition for evaporation of water into the air.

When this same air, with its load of water, meets the cooler walls of the room, its temperature is lowered. Since it can then hold less water than when it was warm, some of its water is condensed against the wall.

EVAPORATION AND SWEATING

NEEDED: Observation.

EXPERIMENT: Open the bathroom door after taking a shower. The air is cold until the body is dry.

REASON: Water, to evaporate, takes energy. Evaporating from the body, water takes energy mainly in the form of body heat. The body actually becomes cooler during the evaporation process. This is one of the reasons we sweat on a hot day.

Water evaporates more quickly from a human body than it does from a table in a 70 degree room because the water on the body is nearer its boiling point.

STUBBORN PAPER

NEEDED: Three strips of paper held together as shown in the drawing, a glass of water.

EXPERIMENT 1: Notice that while the papers are dry they stand apart as shown in the upper drawing. In the water they still stand apart. But when they are brought out of the water they stubbornly cling together.

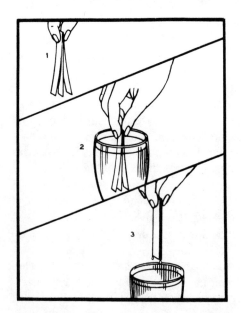

Stubborn Paper

EXPERIMENT 2: A piece of wet paper towel sticks to the smooth top of the kitchen table: adhesion.

EXPERIMENT 3: See if a piece of waxed paper will stick to the smooth table top as does the paper towel. Wet the paper first; it will be difficult to wet. It may be necessary to put water on the table then press the paper into the water to make it stick.

REASON: When the strips are surrounded by large volumes of air or water there is little or no effect on them due to surface tension. The tension is the same on all sides of the papers when they are immersed. But when they are lifted from the water the water films remaining on them are pulled by their surface tension into the least possible space, and the least space is when the strips are made to cling together.

STUCK STICKS

NEEDED: Three matches with square sticks, some water.

EXPERIMENT 1: Challenge a friend to lift two matches with one. It is easy if the sticks are wet; the two matches may be lifted as shown in the lower drawing.

EXPERIMENT 2: Most grains of cereal will stick together when in milk. This is more adhesion.

REASON: Molecules of water cling together: this is called cohesion. They also cling to the match sticks: this is called adhesion. The adhesive force of the sticks and water is great enough to support the weight of the sticks.

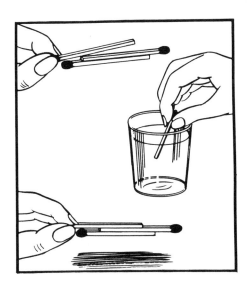

Stuck Sticks

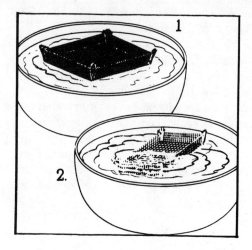

Boat of Holes

BOAT OF HOLES

NEEDED: A piece of new screen wire, a dish of water, some detergent.

EXPERIMENT 1: Bend the wire mesh into the form of a flat bottom boat. Place it carefully, flat, on the surface of water. It floats.

EXPERIMENT 2: Push a side or corner of the wire boat under, and the whole boat sinks rapidly. If part of the surface tension film is broken the weight of the boat causes it to break the films at other parts, and the boat sinks.

EXPERIMENT 3: Float the boat again, and pour a little detergent into the water. The detergent breaks the surface film and the boat sinks.

REASON: Surface tension on the water is like a stretched film, and is strong enough that the wires of the boat do not break through easily.

WATER STREAM FROM A FAUCET

NEEDED: Observation

EXPERIMENT: Turn on the water in the sink. If the flow is right the stream starts the size of the faucet opening, then grows smaller as it falls.

REASON: Surface tension tends to hold the water in a solid stream. But as it falls its speed increases because of the pull of gravity. It is in continuous flow, which means that the amount passing any point must be the same. If it is flowing faster the size must be reduced to meet this law of physics.

The amount of flow is the cross-sectional area of the stream times the speed of flow of the water.

Water Stream from a Faucet

CAPILLARITY

NEEDED: Strips of paper towel, salt water, soapy water, plain water, rubbing alcohol, oil.

EXPERIMENT: Dip the ends of the paper strips into the different liquids, and notice that the liquids do not move up into the paper, between the fibers, to the same extent.

REASON: "Sometimes the forces of cohesion and adhesion oppose each other; this brings about the effect known as capillarity." (The *Book of Popular Science* explanation).

Capillarity is a surface tension effect. The small openings in the paper towel are capillary openings. The molecules of the different materials and the dissolved materials in water have a varied effect because of variation in their surface tension.

Capillarity

The Fisherman's Puzzle

THE FISHERMAN'S PUZZLE

NEEDED: A jar of water, a can, nails, a marking crayon.

EXPERIMENT: A fisherman in a boat on a very small lake throws some heavy iron out of the boat and into the water. Does the act raise or lower the level of the water in the lake?

Let the jar represent the lake, the can the boat. Put nails in the can and float it on the water in the jar. Mark the water level. Now take the nails out of the can and drop them into the water. Replace the can so it floats again. Note that the water level in the jar is lower.

REASON: In the boat or the can the iron (nails usually are iron) displaces a weight of water equal to its weight because they are floating with the boat. In the water the iron displaces only an amount of water equal to its volume. Since the iron is much heavier than water its volume is much less than the volume of water equal to its weight.

FUNNEL FALLING

NEEDED: Dry sand, sugar, salt or other grains or crystals, funnel.

EXPERIMENT: Fill the funnel, then watch as the finger is removed and the grains "pour" out as if they were liquid.

Funnel Falling

COMMENT: A liquid can be described as a substance that, unlike a solid, flows readily, but unlike a gas, does not expand indefinitely. The grains or crystals fit this description.

The middle grains fall first, making a cone-shaped design like water. But, unlike water, they do not begin to swirl as they fall through the funnel.

WATER BULLETS

NEEDED: A pan of water, a thermometer, a fan.

EXPERIMENT 1: Place water in the pan. We know it will evaporate if left alone, and will evaporate faster if air from the fan blows over it. Place the thermometer in the water, and it will be cooler when the fan blows air over it.

REASON: Molecules of the water are in constant motion, as are molecules in any fluid. Some of them gain a high enough speed to shoot like bullets out of the water into the air above. Many fall back into the water; some escape. More can escape if a breeze is blowing to blow them away.

EXPERIMENT 2: Why does hot water evaporate faster than cold water? The molecules move faster and more of them escape the solution.

Water Bullets

Evaporation is an example of expansion, and expansion means cooling. In this case the water molecules escaping carry away some of the kinetic energy of the water, and a lowering of the kinetic energy means a lowering of the temperature.

AN EVAPORATION TRICK

NEEDED: Two similar jars, rubber from a large broken balloon, water, rubber bands.

EXPERIMENT: Fill one jar almost full of water. Set both jars aside in a warm room for a few hours, until the temperature in each jar is the same. Cover both jars with stretched pieces of rubber held on tightly with rubber bands, sealing the jars.

An Evaporation Trick

The rubber over the jar of water will be seen to bulge slightly, showing that a little pressure has been generated in the small air space above the water.

REASON: Molecules of water, always moving, sometimes jump out of the water and into the air. This is evaporation, and can increase the pressure in the air enough to bulge the rubber. The jar of air has nothing to evaporate and so the pressure remains the same as long as temperature in the jars remains the same.

The bulge is somewhat exaggerated in the drawing to make for clarity. The jars should remain several hours or overnight after being sealed with the rubber, to show the bulge.

BLOWING SNOW

NEEDED: Observation of light snow in the wind.

EXPERIMENT 1: Observe the snow on a windy day. When the wind is strong the snow seems to blow away, while the temperature is below freezing and we know it cannot melt.

EXPERIMENT 2: Leave some moth balls out in the open. They will disappear or "sublime." This takes a little time.

REASON: Snow is made of ice crystals, and ice is one of the substances that can pass from the solid to the gaseous state without going through the liquid state. This is called "sublimation."

So, we see that the snow sublimes as the wind blows it about—without melting. Large pieces of ice do not usually do this to a noticeable extent, however.

Blowing Snow

Mystery of the Melting Ice

MYSTERY OF THE MELTING ICE

NEEDED: A glass of water and one or two large ice cubes.

EXPERIMENT: Place the ice in the glass, then fill the glass with water just to the overflowing point. Since the ice extends above the water surface, an overflow is expected as it melts. This does not occur.

REASON: Ice is one of the substances that expands when freezing. It displaces exactly its weight in water as it floats. As it melts its volume is reduced to about 9/10 of its original volume, and this smaller volume does not overflow the glass.

When the ice formed it expanded to occupy almost 10/9 of the volume it had in the liquid state, but it did not increase in weight.

If the water is allowed to get warm after the ice melts it should overflow, because it will expand as it warms to room temperature.

MYSTERIOUS ICE CUBES

NEEDED: Two salad bowls or small baking dishes, a pitcher of ice cubes in water.

EXPERIMENT 1: Fill a bowl almost full of water from the pitcher of ice water. Fill another bowl equally full of warm water. Try to float an ice cube in the center of each bowl.

EXPERIMENT 2: Float two ice cubes in the bowl. Do they move apart or come together?

REASON: The ice cube may be floated in the center of the ice water without difficulty after the water has stopped moving. But it should be impossible for the cube to remain in the center of the warm water.

This is because the warm water melts the ice, the cooler water from the melting ice goes downward, causing irregular convection currents that rise and move to the edges of the bowl. The moving convection currents carry the ice with them to the edge.

Mysterious Ice Cubes

WATER IN THE FLAME

NEEDED: A flame as from a candle or gas.

EXPERIMENT 1: Hold a container of cold water above the flame briefly, and water will be seen condensing on the cold container surface.

EXPERIMENT 2: Note the white "smoke" coming from an automobile exhaust on a cold day. The white particles are water droplets. Burning of the fuel in the engine produces water, carbon dioxide, carbon monoxide, and other gases.

ADHESION, COHESION

NEEDED: Two squares of wood suspended from a yard stick as shown, water, a pan for the water, sand.

Water in the Flame

EXPERIMENT: Let the square of wood which is suspended over the water touch it. Pick it up from the water, then balance the squares by putting sand on the other lighter one.

Adhesion, Cohesion

Now let the wet square come down to the water surface. It will cling to the water surface, and a surprising amount of sand will have to be put on the other square to pull it loose.

REASON: The attraction between different kinds of molecules (here water, and wood) is called adhesion. The attraction between molecules of the same kind (here, water) is called cohesion. When the square is pulled away from the water it will be wet on the bottom. It pulled some water up with it. This shows that the adhesion between wood and water is greater than the cohesion between the molecules of water.

THE REYNOLDS RIDGE

NEEDED: A piece of glass, running water, soap.

EXPERIMENT: Clean the glass thoroughly, then hold it under the faucet until the soap has been washed off. Take it away from the water, then before the film of water has drained off, touch the bottom of the glass with a soapy finger. A ridge will be seen to rise up the glass in the water film.

REASON: The clean water has greater surface tension than the water contaiminated by the finger at the bottom of the glass. The greater surface tension draws the contaminated water up, making a ridge where the two kinds of water meet. This is called the Reynolds

The Reynolds Ridge

Ridge. It may be seen many places in nature, particularly at the edges of bodies of water.

Oil on the finger tip will cause the ridge to move up, but more slowly. The finger alone usually will contaminate the water enough to case a slowly-moving ridge. To make a standing ridge in a natural body of water both water and contaminant must be flowing. The ridge moves upstream exactly as fast as the water moves downstream. This can often be seen at a stream bank.

THE HYDRAULIC JUMP

NEEDED: A cookie sheet and the kitchen sink.

EXPERIMENT 1: Let the water pour onto the sheet, and the movement of the water will form one type of hydraulic jump, as shown in the illustration.

EXPERIMENT 2: Watch the water spilling over a dam. It flows rather smoothly to the "jump," then froths, foams, and bubbles in turbulence at the point where its speed decreases.

REASON: The hydraulic jump is a stationary wave very complicated and difficult to analyze. Higher mathematics is required.

The hydraulic jump is a sort of shock wave, similar to the sonic boom emitted by an object traveling faster than sound.

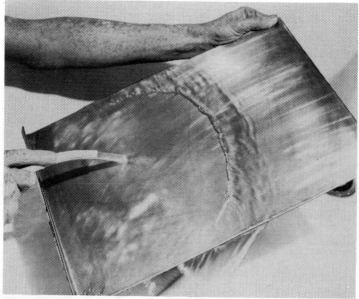

The Hydraulic Jump

The jump is actually a wave travelling upstream at the characteristic velocity of a wave on the water. It is a standing wave because the wave velocity at the point and the water velocity are the same. The wave is traveling upstream at its characteristic velocity relative to the moving water, but it is standing still relative to the land.

SYPHON WITHOUT SUCKING

NEEDED: Two containers, water, a small hose.

EXPERIMENT: Place water in one container; set it on a table. Place the other container on a chair or floor so it is lower than the first.

Hold up the hose, and let it coil into the water so that it fills with water. Place a finger over the end and lower that end into the other container. Remove the finger and the syphon starts.

REASON: Placing the finger over the end of the hose allows the hose to remain full of water as it is lifted over the edge of the container. When the end of the hose is placed lower than the water surface the pressure of the atmosphere on that surface forces water to flow down the hose.

Why doesn't atmospheric pressure act equally on the lower end of the hose? Because the pressure in the lower end of the hose is at

Syphon Without Sucking

atmospheric pressure plus pressure of the column of water in the hose that acts against it.

Don't suck on a hose to start a syphon. Some liquids can be dangerous if sucked into the mouth.

SHEET EROSION

NEEDED: A pan of top soil, some coins or flat stones, rain or a soft sprinkling from a hose.

EXPERIMENT: Place the coins or stones on the soil in the pan, place the pan out in the rain, and watch.

Soil is slowly washed away, and where the coins or stones give protection the erosion is not as noticeable. This type of erosion, hardly noticed, gradually carries away the topsoil from a field. It is called sheet erosion. The best remedy is to keep something growing on the soil at all times. Grass protects the soil from the pounding of the raindrops, and the roots bind the soil together, so that there is little or no erosion.

Sheet Erosion

WATER SPURT

NEEDED: A plastic vinegar jug with cap, water.

EXPERIMENT: Make a small hole near the bottom of the jug. Fill the jug with water, screw the cap on, and the water does not spurt out. Loosen the cap and the water spurts out again.

REASON: This is a matter of surface tension. The weight of the water normally would make the water spurt out, but when no air gets in at the top of the jug the water does not flow because the pressure of the atmosphere against the water at the hole—and the surface tension—keeps the water in.

If the hole is large, air can get in through it at the same time water spurts out, and the water will gurgle out. But if the hole is small, surface tension of the water at the hole is strong enough to prevent the flow of water out and air in at the same time.

Water Spurt

CAPILLARY ATTRACTION

NEEDED: Eye-dropper with a long thin snout, water, a glass.

EXPERIMENT: Note that the water in a glass curves upward where it touches the glass. Now touch the end of the dropper to the water surface. The water not only rises where it touches, but moves slightly upward into the dropper.

Capillary Attraction

REASON: The same thing happens in both cases, but the small end of the dropper allows the surface tension to pull up a little water not only at the edge of the container but throughout the container's width.

A small glass tube is called a capillary tube, from a Greek word meaning "hair." The smaller the tube the higher the capillary action will lift the water. A cloth or piece of rough paper touched to the water surface shows capillary action; it is made up of small fibers that make the water rise between them.

CLINGING WATER

NEEDED: A saucepan, a bowl of larger diameter, a water hose.

EXPERIMENT: Pour water into the bowl. When full, the water will flow down the sides of the bowl until the pan is full, then will flow down the sides of the pan. The water clings to the bowl and pan, some of it flowing out on the handle of the pan before flowing to the sink or ground.

REASON: Adhesion, here, is attraction between unlike molecules, those of the water and those of the objects. The attraction causes the water to cling to the pan and bowl. Cohesion causes the water molecules to cling together.

There are many interesting and simple surface tension adhesion and cohesion experiments. Molecules of water in a container cling together so that the water surface will support small steel objects such as needles and razor blades. This is an example of surface tension experiments.

Clinging Water

THE DRIP

NEEDED: A water faucet (spigot or tap mean the same) or an eye dropper. Water.

EXPERIMENT 1: Open the spigot a little; water will pour downward in a narrow stream. Close the spigot slowly; a point will be reached where the water will not flow, but will drip.

EXPERIMENT 2: The same experiment may be performed with an eye dropper.

When the flow is decreased in the right amount, the surface tension squeezes the water into droplets as they leave the spigot surface. The drawing may help to show this.

SUDS

NEEDED: Two jars, detergent, the kitchen sink.

EXPERIMENT 1: Put a half teaspoonful of detergent in both jars. Let water run slowly into one; let it run fast into the other. The second jar will fill rapidly with suds that overflow.

The Drip

EXPERIMENT 2: Try this with hot water.

REASON: The rapidly moving water breaks through the surface of the water in the jar, carrying air with it. This air "blows bubbles" which increase the volume of liquid in the jar and also decrease its density. As more water comes in and more bubbles are formed they float on the surface of the water, rise, and some of them overflow.

The slow-moving water in the other jar carries very little air with it into the detergent solution, and so does not blow as many bubbles.

Suds

WATER CAVITIES

NEEDED: Water in a tub or pool, a large spoon.

EXPERIMENT: Move the hand or spoon quickly through the water. If it is moved fast enough, a bubble resembling air will be seen behind the hand or spoon.

REASON: Pressure behind the moving spoon or hand is reduced—a partial vacuum is produced, causing the water to go through the same phenomenon as boiling. Water can boil from heat, also it can boil from reduced pressure. Boiling occurs when the vapor pressure of the water is equal to the surrounding pressure.

This effect, called "cavitation," is damaging to propellers on large boats, and to large metal tubes through which water flows, such as spillway tunnels at dams. In clear water, cavitation can be seen clearly in the wake of boat propellers.

In the drawing the spoon is being moved in the direction the arrow points.

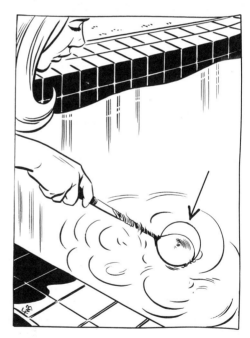

Water Cavities

LIQUID LAYERS

NEEDED: A tall glass, water, alcohol (rubbing alcohol will do), cooking oil, syrups of different thicknesses.

EXPERIMENT: Fill the glass half full of water. Add about a third that much alcohol—the water and alcohol will mix. Add a little of

Liquid Layers

the thickest syrup, then a little of the thinner syrup. They will sink to the bottom of the glass and remain in layers. Finally, add cooking oil; it will float on top.

REASON: Specific gravity is the weight of a substance compared with the weight of an equal volume of water. Liquids with specific gravities higher than that of water, which is 1, will sink in water; liquids with specific gravities lower than that of water will float on water unless they mix with it, as did the alcohol.

Alcohol has a lower specific gravity than water, so when it mixes with water the specific gravity of the mixture is lower than 1. From the bottom upward the lighter liquid layer floats on the next heavier layer below it if they do not mix or dissolve in one another.

SAND CASTLES

NEEDED: A sandy beach.

EXPERIMENT: Build a sand castle, and note how the sand particles hold together when wet yet fall apart when dry.

REASON: Water molecules pull to each other in all directions when there is water in all directions. When there is water only on three sides, however, the pull is much like a stretched rubber sheet. We call this surface tension. It is this surface tension that holds the wet sand particles together.

If the sand is dry there can be no surface tension. Sand under water does not have a surface tension effect because the attraction between molecules is equal in all directions. The sand castle can be built when all grains of the sand are attached to all other grains by the surface tension of the water between them.

Fluids are usually considered to be either gases or liquids. Yet other substances act like fluids, including dry sand, flour, or even gravel as it is poured from a truck.

(Suggested by Pierre LaFrance, in the magazine *The Physics Teacher,* January 1975.)

Sand Castles

DRY SOLIDS—CONTACT ANGLES

NEEDED: A jar of wet sand, a jar of dry sand.

EXPERIMENT: Pour a little water into each jar. The water will sink quickly through the wet sand, more slowly through the dry sand.

REASON: A drop of water on a solid surface forms a "contact angle" which can be zero, in which case the water does not wet the solid at all. If the contact angle is 180 degrees the liquid wets the solid perfectly. If the solid surface has been wet, then we must think of the contact angle of the fluid with itself, which is 180 degrees.

183

Dry Solids—Contact Angles

In the dry sand the water forms little arches across the grains at the characteristic contact angle of water with silicon dioxide or sand. When the sand has been previously wetted the water runs through freely, following the films already established by the wetting.

Note: There is a resistance to wetting which requires a small amount of energy to overcome.

RISING WATER

NEEDED: A glass of water, a piece of paper towel.

EXPERIMENT 1: Sight along the surface of the water in its glass. It is a little higher where it meets the glass than elsewhere. Adhesion is the force that holds the water to the glass, causing the water to rise against the surface of the glass.

EXPERIMENT 2: Dip the end of a strip of paper towel into the water. The water will rise through the fibers of the paper, higher than it did at the edge of the glass, but for the same reason. The rise of water in the narrow spaces between the paper fibers is called capillarity, from the Latin word capillus, meaning hair.

REASON: Movement of the fluid in capillarity is not always up. It may be down. If a narrow glass tube is pushed into mercury the part of the mercury next to the glass is depressed. Glass attracts water more than water attracts water; mercury attracts itself more than it is attracted by the glass.

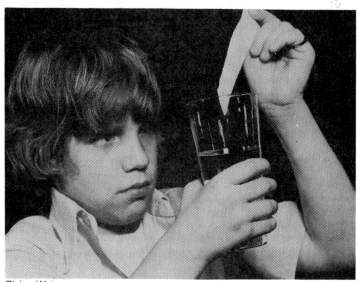

Rising Water

CLEAR ICE CUBES

NEEDED: Water, ice trays.

EXPERIMENT 1: Fill one ice tray with water from the spigot. Fill the other with water that is almost boiling hot. Place both trays in the freezer. Ice cubes made with hot water should be clear.

EXPERIMENT 2: Put the tap water into a jar and shake it vigorously, then put it into an ice tray. Shaking should put more air into the water.

EXPERIMENT 3: Let a jar of hot water cool, then shake it. The water then should make cubes full of bubbles. Shaking dissolves air into the water.

REASON: All water has some air dissolved in it. Water direct from the spigot has a considerable amount; boiling or heating releases most of the dissolved air, which comes off as bubbles. Freezing also releases dissolved air, but instead of bubbling off, it stays in the cooling water, allowing ice to freeze around it. It is this trapped air in ice cubes that accounts for most of the cloudy or streaked appearance of the cubes.

Clear Ice Cubes

A WATER UMBRELLA

NEEDED: A water hose, a jar with a smooth lid.

EXPERIMENT 1: Direct the water stream onto the jar lid. With practice the water can be made to form an "umbrella" as shown.

A Water Umbrella

EXPERIMENT 2: Hold a spoon under the flow from the kitchen sink faucet. The umbrella can be formed.

REASON: Surface tension keeps the water in a sheet as it flows over the edge of the jar top. As the sheet gets thinner and falls at a greater speed the surface tension pulls it into drops.

Surface tension acts as a rubber sheet on the surface of water, tending to draw it into the shape having the smaller surface area.

SWELLING FRUIT

NEEDED: Dried fruit, water, salt, two bowls.

EXPERIMENT 1: Put salt water in one bowl and plain water in the other. Place dried fruit: apricots, prunes, apples for example, in both bowls. In 12 hours the fruit in the plain water will swell. Fruit in the salt water will not.

EXPERIMENT 2: Salt is a preservative. Put small cucumbers in strong salt solutions; the living cells that may lead to decay are dehydrated as the salt solution draws their liquids out through osmosis. The cells cannot live.

Swelling Fruits

REASON: In osmosis water flows both ways through cell membranes. But the flow is greater toward the more concentrated solution, which in this case is inside the dried fruit. The flow of water into the fruit causes it to swell.

If salt water is tried there is likely to be no flow, since the concentration inside and outside of the cells of fruit should be about the same.

CLOUD IN A JUG

NEEDED: A milk jug and a match

EXPERIMENT: Hold the lighted match under the mouth of the jug for a few seconds, then blow hard into the jug, compressing the air in it as much as possible. Release the air pressure suddenly, and a cloud will appear in the jug and remain for an instant.

REASON: Blowing into the jug compresses the air and adds a little moisture from the breath. The compression heats the air slightly, but most of the heat is absorbed by the glass walls of the jug. As the pressure is released, the expansion of the air cools it, and since it cannot then hold as much moisture as the warmer air, some of the moisture is condensed briefly into droplets that make up the cloud.

The lighted match warms the air in the jug slightly, and adds tiny particles of smoke around which water vapor molecules can condense to form the fog. Putting particles into air for this purpose is called "seeding" the air. The invisible separate molecules of water

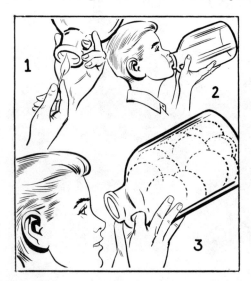

Cloud In a Jug

vapor condense when cooled on the solid particles to make the visible fog.

Repeat the act of compressing and releasing the pressure in the jug. The cloud becomes more dense each time if this is done quickly.

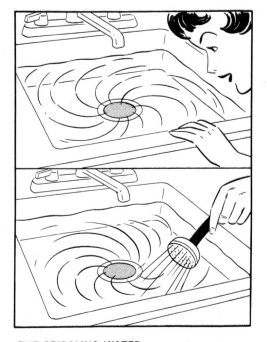

Spiraling Water

THE SPIRALING WATER

EXPERIMENT: Notice the direction of the spiraling of the water as it goes down the drain. Contrary to the statement sometimes heard that it always whirls clockwise in the southern hemisphere and counterclockwise in the northern hemisphere, it can be seen to whirl either way.

The turning of the earth on its axis can start the spin in agreement with the statement, provided there is no motion in the water to begin with. But a very slight motion in the water can determine which way it will whirl. The vortex seen is due to gyroscopic action in the water. Look up "Coriolis Effect."

PHYSICAL EQUILIBRIUM

NEEDED: A large jar, a little hot water, a cover, perferably glass, for the jar.

EXPERIMENT: Cover the jar with the hot water inside, and watch what happens. The water tends to evaporate, yet the vapor condenses on the sides of the jar and the top, changing back into water. Physical equilibrium is defined as that situation in which the liquid molecules evaporating and vapor molecules condensing are equal in number.

"MAGNETIC" SUGAR

NEEDED: A bowl of water, a wooden match or toothpick, a sugar cube.

EXPERIMENT: Float the wood on the water. Dip a corner of the sugar cube into the water near the end of the wood, hold it there, and the wood will begin to move toward the sugar. Move the sugar away slowly, still keeping it in the water, and the wood will follow it as if drawn by magnetic force.

REASON: There is no magnetism involved, of course. The dissolving sugar makes the water heavier at that point, the heavier water flows downward, causing a current in the water. The wood is moved by the water that flows in to take the place of that which has gone down.

If the cube is touched lightly to the water just in front of the wood, the wood is likely to move away. This is because the sugar beginning to dissolve in the water reduces the surface tension at that end of the wood. The surface tension at the other end is as strong as before, and like a stretched rubber sheet, pulls the wood back. But as more sugar dissolves and begins to fall through the water, the flow begins that will move the wood toward the sugar cube.

SMOKE RING IN WATER

NEEDED: A dish or jar of water, a drop of ink or colored water.

EXPERIMENT: Drop the ink into the water, and in many cases it will form into rings resembling smoke rings.

REASON: As in the formation of vortex rings with the Smoke Cannon, there is a drag where the outer edges of the ink drop move through the water, causing the ink to form a toroidal or doughnut shape. Clear water pulled into the middle by the turning effect keeps the center of the doughnut more or less clear.

THE CROOKED WATER SURFACE

NEEDED: A glass of water and a small lid that will float.

EXPERIMENT: Fill the glass almost full. Float the lid on the surface of the water, and it will move to the edge of the glass. Fill the glass completely, and the lid will float in the center of the water surface.

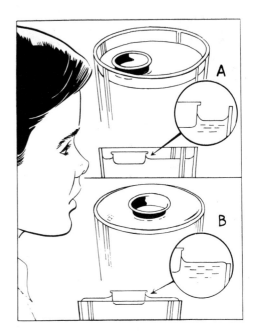

The Crooked Water Surface

REASON: Notice that when the glass is not filled, the water surface between the lid and the glass is bent upward as in diagram A. The surface tension of the water, acting like a stretched rubber sheet, draws the lid to the nearest part of the glass.

When the glass is filled, the water surface is in the shape shown in diagram B. Here the surface tension acts to push the lid away from the glass, with a similar effect at all points on the glass.

WET MUD

NEEDED: Observation only. We know that mud will not brush off of our clothes, but will brush off after it dries.

REASON: The surface tension of water binds the clay particles into mud. Without this bond, the dry loose particles easily brush apart. The clay particles are also bound to the cloth by surface tension.

SURFACE TENSION

NEEDED: Two dishes of water, two ping-pong balls, a small piece of soap.

EXPERIMENT: Place a ball on the surface of the water in a dish, and it will move to the nearer side and cling to the dish. Mix some soap in the water in the other dish, place the ball on the water surface, and it will remain in the center of the dish.

Surface Tension

REASON: The water molecules attract the ball and the dish. The effect, surface tension, can be seen as the curved shape of the water surface. The double effect of the surface tension causes a small force to pull the ball to the nearest side of the dish. Soap in the water greatly weakens the surface tension, so that the curving effects are not noticed. The water surface will appear flat, and there will be insufficient force to pull the ball to a side.

THE "LIVE" WIRE

NEEDED: A piece of fine wire such as that found in lamp cords, a bowl of water, a drop of oil.

EXPERIMENT: Wind the wire into a flat spiral shape as shown, place it on the water surface carefully so it will float, and let the water become still. Place a drop of oil in the center of the spiral, and the wire will "come alive" and begin to turn.

REASON: The oil floats on the surface of the water and tends to spread out evenly over that surface, reducing the surface tension beginning at the center and spreading outward. There is a slight force on the wire in the direction of the arrows, and this makes the spiral turn.

THE RISING BUBBLE

NEEDED: Soap solution and a glass or plastic funnel.

EXPERIMENT: Dip the funnel into the soap solution and withdraw it, so that a film of soapy water is lifted with it. The film will be at the large end of the funnel, but will crawl slowly up to the little end as the funnel is held as shown. (The funnel must be clean and wet with soapy water for this to work successfully.)

REASON: The film has two surfaces, both of which act like stretched rubber sheets (surface tension is the term for this). They tend to shrink to the smallest surface area, and their force due to surface tension is great enough to lift the film upward against the pull of gravity. Dirt tends to reduce the adhesive forces due to surface tension.

The Rising Bubble

SURFACE TENSION

NEEDED: Two toothpicks, a bowl of water, a small piece of soap.

EXPERIMENT: Place the toothpicks side by side on the still water surface, and they tend to move together. Touch the soap to the water between them, and the toothpicks move apart.

REASON: Soap and many other substances will reduce the surface tension. If the soap is placed between the toothpicks the surface tension there is reduced, allowing the greater surface tension effect on the outside of the toothpicks to pull them farther apart.

Dip a toothpick into alcohol, and touch it to the water surface between the floating toothpicks. The floating toothpicks will spring apart for the same reason. Alcohol, too, reduces the surface tension.

A Surface Tension Puzzle

A SURFACE TENSION PUZZLE

NEEDED: A bowl of water, an ice cube, two matches or toothpicks.

EXPERIMENT: Let the water become still, then place the matches side by side as shown. Touch the ice cube to the water between the matches. The cooling of the water increases the surface tension and should bring the pieces of wood together. But it does not work this way. The matches move apart.

REASON: Water currents set up by the melting ice float the matches apart. The effect of these currents is stronger than the pull due to the increased surface tension of the slightly cooled water. The water currents move both downward and laterally.

HOLES THAT HOLD

NEEDED: Small mesh kitchen strainer, water, oil.

EXPERIMENT: Pour water down the side of the strainer, and it runs through. Dip the strainer into oil, sling out the excess, then try the water. The water will collect in the bottom, and not run through the mesh until a considerable amount has collected in the bottom of the strainer.

REASON: The surface tension or molecular attraction is strong enough between the meshes of the wire to hold water. If the wire is not oily the water can "wet" it and flow easily on its surface to the underside, then spill from the strainer.

While the strainer is holding water, touch a finger to the underside of the mesh. This breaks the surface tension and allows the water to flow through to empty the strainer.

Holes That Hold

THE MOVING TOOTHPICK

NEEDED: A toothpick, some soap and some water.

EXPERIMENT: Stick one end of the toothpick into the soap, pull it out, then float the toothpick carefully on the water surface. It moves about mysteriously.

REASON: The soap decreases the surface tension of the water at that end of the toothpick, and the stronger pull due to greater surface tension at the other end causes the toothpick to move along. The water surface must be calm.

BUOYANCY

NEEDED: A tin can with both ends cut out, a cardboard square, a bucket of water.

EXPERIMENT: Completely cover one end of the can with the card, and push the can and card down into the water. The buoyancy

Buoyancy

of the water will keep the card against the end of the can. Then slowly pour water into the can. The card stays on until the water inside the can reaches the height of the water outside it, in the bucket, or until the can and card are completely submerged. Then the card falls off.

REASON: Pressure beneath the surface of water at any particular depth is equal in all directions, and when there is no water in the can the water pushes against the bottom of the card to hold it in place. As water is poured into the can the pressure on the bottom of the card decreases, and when the water inside reaches the level of water outside the can there is no pressure on the bottom of the card in excess of that above. The card falls off in response to the pull of gravity.

Chapter 7
Gravity & Centrifugal Force

A SAND AND WATER MYSTERY

NEEDED: Sand, water, a glass

EXPERIMENT: Stir sand and water in a glass, and the sand will be seen to form a pile in the center as shown.

REASON: Because of the friction between the glass and the water, the bottom section of the water moves more slowly; therefore less centrifugal force is exerted. This causes the water to flow downward along the sides of the glass, inward across the bottom.

The upward flow of the water is not sufficient to carry the sand upward along the complete water path, but just enough to pile it up on the bottom of the glass.

MORE SAND AND WATER MYSTERY

NEEDED: A glass of water, some sand, some string.

EXPERIMENT: (1) Stir the water and sand together, and the sand will pile up in the middle of the glass as the water turns around. (2) Suspend the glass by the string, twist the string and release it so that the glass with the water will whirl around. This time the sand will go outward along the bottom of the glass.

REASON: (1) Friction between water and glass at the bottom is greater than the friction higher up in the glass; therefore the water at the bottom turns more slowly. This causes downward currents along the sides of the glass, inward at the bottom, and upward in the center. The water motion carries the sand to the center.

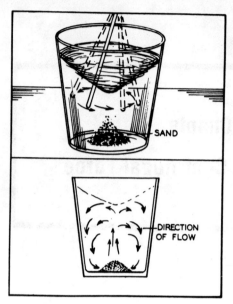

A Sand and Water Mystery

(2) When the glass, too, turns with the water and sand, all turn at the same rate, so that there is no greater friction at the bottom than higher up. Therefore, there is no flow of water as in (1). The sand can then follow its natural tendency to move off at a tangent to the circular motion, which means that it can move to the rim of the glass as shown.

More Sand and Water Mystery

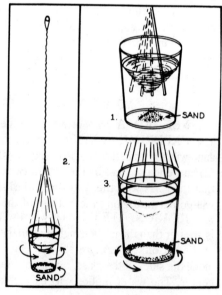

SHAPE OF THE POURED WATER

NEEDED: A pan of water.

EXPERIMENT 1: Pour the water out of the pan, and notice that the stream becomes round as it flows down a few inches from the pan.

EXPERIMENT 2: Hold the pan higher, and the stream will tend to divide into drops.

REASON: Gravity gives the water in the pan a level surface, and as it pours out of the pan, it takes the form shown in the diagram at the top.

As it pours down, gravity and the shape of the pan influence it less and less, until surface tension pulls the stream into the shape shown in the lower cross-section drawing.

Surface tension is a force which tends to give water and other liquids the shape with the smallest surface area.

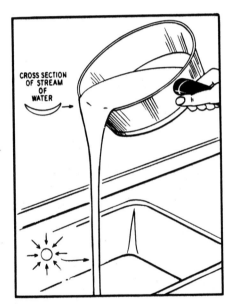

Shape of the Poured Water

WHY THE VORTEX?

NEEDED: A round pan or jar, some water, something to stir with.

EXPERIMENT 1: Stir the water, and the faster it goes around the deeper will be the "hole" in the center. This hole with the rotating water around it is called a vortex.

EXPERIMENT 2: Use something larger to stir with, such as a spoon. The water may be made to reach the top of the container and flow over the rim.

REASON: Centrifugal force supplied by the stirrer tends to cause the water to fly away from the center. The wall of the vessel will not let the water fly off.

The pressure under the water increases with depth. At the bottom of the vortex the force due to pressure will not allow the water to move out near the side of the glass as it does near the top.

WEIGHTY PROBLEMS

NEEDED: A scale, a heavy weight.

EXPERIMENT 1: Stand on the scale and raise one foot. Do you weigh less?

EXPERIMENT 2: Lift the weight with one hand, then with two. It seems much lighter when lifted with two hands.

REASON: (1) When you stand on two feet, your weight is applied to the scale platform through both. Lift one foot, and its weight is transferred to the other, and through it to the scale. The reading is the same.

(2) When the weight is lifted with two hands, each hand has to support only half as much as if it is lifted with one hand. The weight is the same, but feels less when two hands divide it. This is simply a psychological illusion.

A Bird Mystery

A BIRD MYSTERY

NEEDED: A wooden box with a perch, a bird, a scale.

EXPERIMENT 1: Watch the weight scale. There will be some variation as the bird leaves the perch and returns to it, but the *average* weight of the bird on the perch and in the air inside the box will be the same.

EXPERIMENT 2: In an open cage, the bird's weight in flight may register less because part of the downward movement of air caused by the flying will go outside the cage and will not register on the solid bottom.

REASON: When the bird is flying, its weight is held by the air, and the downward push is transmitted through the air to the bottom of the box. Here, too, there will be some variation in the weight recorded, but the average will be the same as when the bird is at rest.

Chapter 8
Electricity & Magnetism

A POTATO TEST

NEEDED: A battery, a potato, some wire.

EXPERIMENT: Connect wires to the battery. Cut the potato in half, and stick the ends of the wires into the potato about an inch apart. A green color will appear in the potato around the positive wire, and bubbles will come from the potato where the negative wire is inserted.

REASON: The green color is due to partial ionization of the copper as negative ions from the solution in the potato are neutralized at the anode pole and attack the copper to form an ionic copper salt. This action is very slight since the concentration of ions in the potato juice are very low. The bubbles of gas are due to liberation of tiny volumes of hydrogen gas at the negative pole.

This experiment has been suggested to determine which of the two wires is positive and which is negative.

Even if an inert wire, such as platinum, is used, the potato will still show color around the positive wire. The color will be a pale pink, or dirty pale pink. This is caused by oxidization of compounds in the potato by oxygen, or other oxidizing substances such as chlorine released by the electrochemical reaction.

It is the same reaction that causes a cut potato to brown if it is left in the air (apples and peaches will brown the same way if exposed to air or other oxidizing conditions after they are cut).

When preparing things that brown for deep freezing it is customary to sprinkle them with ascorbic acid, vitamin C, sold under a number of brand names, for the purpose of keeping them from

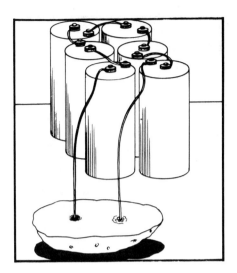

A Potato Test

turning brown. Vitamin C (harmless to the body) is a reducing agent; that is, it reacts with, and takes up, oxygen easier than the materials in the fruit which would turn brown on oxidation.

(Such oxidation destroys the vitamin properties of ascorbic acid, but it does protect the fruit from browning.) Lemon juice contains ascorbic acid and will protect a cut potato or fruit from browning in the air.

EXPERIMENT 2: Put a drop of lemon juice on the potato at the point where the positive wire enters and see if it will prevent the formation of a color at that point.

THE GOOFY PING-PONG BALL

NEEDED: A table tennis ball and a hard rubber comb.

EXPERIMENT: Rub the comb briskly on a woolen sleeve or cloth, move it around in circles quickly as shown, and the ball will follow it. It is not necessary to touch the ball with the comb.

REASON: Rubbing places a charge of static electricity on the comb. The uncharged ball is attracted by the charge on the comb in a very mysterious manner.

The rule is: like charges repel, unlike charges attract, and a charged object near an uncharged object induces an opposite charge on the near side of the formerly uncharged object. The induced charge is due to a shift of electrons on the surface of the formerly uncharged object.

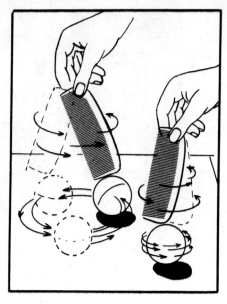

The Goofy Ping-Pong Ball

THE WAYWARD WATER

NEEDED: A hard rubber comb and a thin stream of water from a faucet.

EXPERIMENT: Rub the comb briskly on a woolen cloth or through the hair, then hold it near the stream of water. The water will change its course as it is attracted by the comb.

REASON: Rubbing the comb puts a charge of static electricity on it. The water is not charged until close proximity of the comb repels electrons to the opposite side of the water column. Then the charge is opposite, and the comb and water are attracted to each other. Note that the rule in electricity, "unlike charges attract one another," can be expanded to include attraction between a charged body and a previously uncharged body which has an opposite charge induced in it when close to a charged body. This experiment works only when the air is dry. It will not work in a room where the humidity is high. Much water vapor in air allows the excess electrons of the charged comb to escape from the comb to invisible water vapor aggregates in air.

MAGNETIC OR NOT?

NEEDED: A magnet, a tin can lid, various substances to be tested such as paper, cloth, screen wire, and aluminum.

EXPERIMENT: Lift the can lid with the magnet (it is not tin—it is iron), as shown in the upper drawing. Then try lifting it with various substances or objects placed between the magnet and the lid.

OBSERVATION: The magnetism will penetrate the thin cloth and paper as if it were not there. Substances containing iron will "short circuit" a magnetic circuit, although, if the magnet is strong, some of the magnetism will penetrate even a thick iron piece such as pliers in sufficient strength to lift the lid underneath. The pliers here become magnetized while in contact with the permanent magnet.

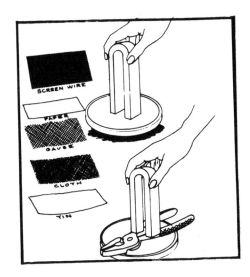

Magnetic or Not?

MONEY POWER

NEEDED: Copper and silver coins, blotting paper, water, a way to measure the current.

EXPERIMENT 1: Stack the coins as shown: dime, blotter, penny, dime, blotter, penny, etc. As long as the blotters are wet, an electric current is produced.

The measuring device for the current shown here is a homemade galvanometer.

EXPERIMENT 2: It consists of a cardboard box wound with many turns of wire. Inside the wire coil a magnetized needle is suspended by a string so it is free to turn. Current in the coil can cause the needle to move. A pocket compass placed inside the coil works well.

Plain water may work in the battery, but it is much better if some sort of electrolyte is mixed with the water. A little salt, or vinegar, or lemon juice will serve.

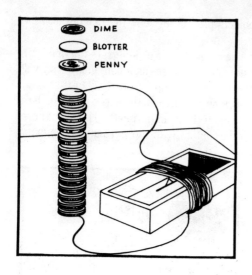

Money Power

THE DIP-NEEDLE

NEEDED: A long needle, a short needle, a cork, a drinking glass, a magnet.

EXPERIMENT: Stick both needles through the cork, and make them balance on the rim of the glass. Then magnetize the small needle by rubbing the eye of it on one end of the permanent magnet. When the needles and cork are again placed on the glass, with the

The Dip-Needle

small needle on a north-south direction, they will not balance level as before.

The needle will dip, to show that the earth's magnetic lines do not run parallel to the surface of the earth, but extend downward toward a point inside the earth. The amount of the dip will vary according to the locality on the earth on which the experiment is tried.

There is no dip at the magnetic equator, which is an irregular line running around the earth, in only a few places close to the geographical equator. Further, there are scattered regions far from the geographical equator where the dip is zero.

At the magnetic poles the dip is 90 degrees. Generally, the dip is small near the equator, and increases (in opposite senses) as you approach the North and South magnetic poles.

TRICKY SWITCHES

NEEDED: A double-pole, double-throw switch, two single-pole, double-throw switches, a battery, a flashlight bulb in a socket, some wire.

EXPERIMENT: Make the connections as shown in the lower drawing, and by changing the positions of the switches, the lamp can be turned either on or off from any of the three switches. (Each switch must be closed one way or the other at all times.) The upper drawing shows how the connections are made in house wiring.

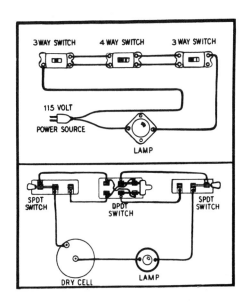

Tricky Switches

The switch in the enter in both cases is a double-pole, double throw switch (abbreviated DPDT) and the switches on both sides are single-pole, double-throw (SPDT) switches.

No boy or girl should use the wiring method requiring line voltage from an electrical outlet—it can be dangerous. The same results are obtained with the lower wiring using a dry cell as the power source and a flashlight bulb as the light.

DIM THE LAMPS

NEEDED: Two flashlight bulbs and a battery to match their voltage, sockets, a double pole, double-throw switch, wire for connections.

EXPERIMENT 1: Connect the lamps as shown. When the switch is closed in one position the lamps burn brightly because they both get the full voltage from the battery. They are connected in "parallel." If the switch is closed the other way the lamps are in "series," in which the same current must pass through one and then the other. Each lamp gets only half the total voltage and therefore is dimmer.

EXPERIMENT 2: It is assumed the two bulbs used in the experiment above are the same rating. Try using one bulb from a two-cell flashlight and one bulb from a three-cell light. If connected

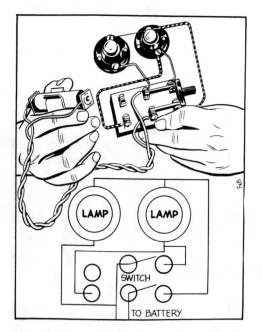

Dim the Lamps

in series one will not burn as brightly as the other; more current can pass through the larger one than the small one can handle without full brightness. The smaller rated bulb will burn brighter than the other.

COMPASS CUT-UPS

NEEDED: Two needles, a ruler, two books, a magnet, some thread.

EXPERIMENT: Magnetize two needles by rubbing the end of a magnet against them as shown in drawing 1, and suspend them on strings so that they will point North and South. Hang them on a ruler. Bring them together by sliding the strings along the ruler, and soon they will point toward each other, not North and South.

REASON: The magnets made by rubbing the needles produce magnetic fields close to them that are stronger than the magnetic field of the earth. The effect of their fields is noticed only when they are close together.

When they hang at a distance from each other, the earth's magnetism keeps them pointing North. The force of attraction of two magnet poles on each other is inversely proportional to the square of the distance between them. The strength of the fields does not vary.

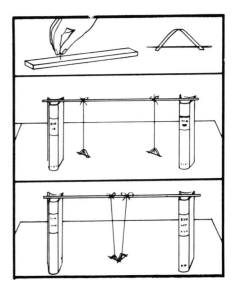

Compass Cut-Ups

THE BASHFUL NEEDLE

NEEDED: Two sewing needles, a magnet, a piece of glass or a smooth table top (not steel or iron).

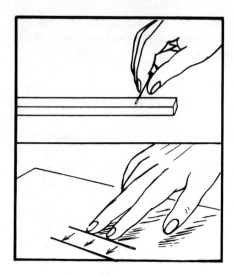

The Bashful Needle

EXPERIMENT: Magnetize the needles by stroking them with one pole of the magnet. Place one on the table, and approach it with the other. If the needles are attracted, turn one around. Then, as one approaches, the other rolls away.

REASON: If the needles approach so that the poles are at opposite ends, they will attract each other. But if they are brought toward each other so that a north pole of one approaches the north of the other, and the two south poles approach each other also, the needles repel each other. (The needle is best stroked flat against the magnet.) The rule is: like poles of a magnet repel; unlike poles attract.

THE BROKEN MAGNET

NEEDED: Two needles, two pairs of pliers, a magnet, a paper clip, a handkerchief.

EXPERIMENT: Magnetize a needle by rubbing it with one end of the permanent magnet. Note that either end of it will be attracted to the other needle.

Wrap the handkerchief around the magnetized needle, grasp it with the pliers, and break it near the middle. (The handkerchief is to prevent pieces of the needle from flying.) You now have two magnets, and their four ends will attract the other needle.

REASON: No matter how many times a magnet is broken, each piece becomes a magnet with a North and South pole. The new magnet will probably be of less strength because of the mechanical jar in breaking.

A SIMPLE ELECTROSCOPE

NEEDED: A jar, a metal wire, a rubber or plastic comb, some Christmas tree icicles.

EXPERIMENT: Bend the wire as shown, so it will hang in the jar, and bend an icicle so it hangs over the wire. Rub the comb vigorously on wool or fur, touch it to the wire, and the foil ends will swing apart.

REASON: The comb becomes charged by friction, and the foil and wire by induction as the comb approaches it. Since both ends of the foil carry the same charge, and since like charges repel, the foil ends push each other apart.

If humidity is high this instrument may not work. Warm it for a while in a warm oven to drive away moisture, then try it again. It should work readily in air-conditioned rooms or almost any heated room in winter.

Christmas tree icicles are made of aluminized Mylar film. The aluminized side must touch the wire. This is better than thin foil for this instrument. The electroscope may be charged by unrolling some types of cellophane tape near the wire rather than by rubbing a comb.

A Simple Electroscope

MAGNETISM BY INDUCTION

NEEDED: A magnet and two nails.

EXPERIMENT 1: Pick up one nail with the magnet as shown in drawing 1, and it will be found that the nail has become a magnet and will pick up another nail. This is magnetism by induction.

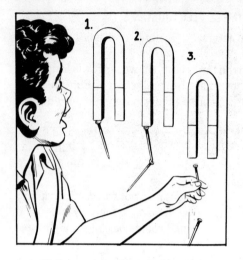

Magnetism by Induction

If the upper nail is pulled away from the magnet, as in drawing 3, the lower nail will fall off, because the upper nail will then have lost practically all of its magnetism.

Soft iron such as nails may be easily magnetized, but loses its magnetism just as easily.

EXPERIMENT 2: Try this with steel needles instead of nails. The bottom needle may not fall off as the top one is moved from the magnet. The top needle may become magnetized through contact with the permanent magnet so that it can hold the lower one. Steel does not lose its magnetism as easily as do softer iron nails.

RUG POWER

NEEDED: A wool rug, a rubber balloon, some bits of newspaper.

EXPERIMENT: Scatter the bits of paper on the floor. Rub the balloon briskly on the rug, hold it above the paper bits, and they will fly up to it.

Some of the paper bits will stick, while some may fly back down again.

REASON: Rubbing the balloon against the wool rug gives it a charge of static electricity which attracts the bits of paper. Some of them may then take up a charge from the balloon, and since it will be the same as that on the balloon, the paper and rubber will repel each other so that the paper flies away (like charges repel; unlike charges attract).

Static electricity experiments do not usually work in summer when the humidity is high. Cold weather outside, warm inside, is best.

Note that the electricity results from the motion the experimenter contributes to the experiment. If no energy is used, no energy is produced.

THE BASHFUL BALLOON

NEEDED: Two balloons, a wool skirt or cloth, a cold, dry day.

EXPERIMENT 1: Place one balloon on a table or floor. Rub the other on the wool until it is thoroughly charged with electricity. Hold it down so that it touches the other.

The second balloon will then be repelled by the first, and may be pushed around rapidly by it without being touched—as if too bashful to touch again.

REASON: When the balloon is rubbed against the wool, it takes on a minus charge. When it touches the second balloon, that, too, becomes negatively charged. They repel each other according to the law that like charges repel and unlike charges attract.

EXPERIMENT 2: A balloon rubbed against wool until it is charged may be placed against the wall. It will remain stuck there; its charge attracts it to the wall, which remains uncharged because of its large area. A charged object is attracted to one uncharged or one carrying a smaller charge.

In these experiments the balloons may be charged usually by being rubbed against a rug on the floor.

EXPERIMENT 3: Rub a plastic or rubber comb through the hair. See if it will attract or repel the balloon.

MAGNETIZE THE SCISSORS

NEEDED: Scissors and a permanent magnet.

EXPERIMENT 1: To magnetize scissors, rub the points on one end of the magnet.

While magnetism is still not perfectly understood, it is believed that in magnetizing a piece of steel, the molecules are caused to align themselves in a more orderly fashion, parallel to one another.

Rubbing the points of the scissors on the other end of the magnet will de-magnetize them, and, unless time is carefully checked, will remagnetize them with reverse polarity.

EXPERIMENT 2: If the points of the scissors are rubbed against the magnet as shown they both become either north or south seeking poles. Try rubbing one point against one pole of the magnet

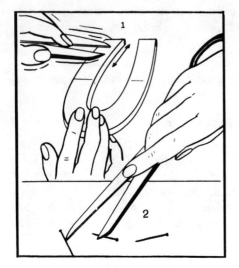

Magnetize the Scissors

and the other against the other pole. The points should then have opposite polarity and should pick up pins better if held separated as in the lower drawing.

A MAGNET MYTH

NEEDED: A strong permanent magnet, two similar pieces of steel such as screwdrivers or needles.

THE MYTH: To make the screwdriver magnetic it must be stroked in one direction on one pole of the strong magnet.

EXPERIMENT 1: Stroke one of the screwdrivers in the recommended way. It will become magnetized. Then rub the point of the other screwdriver back and forth on one pole of the strong magnet. It, too, will become magnetized, proving that the myth is only a myth. The steel need not be stroked in one direction only.

A Magnet Myth

A small magnet and sewing needles may be used to prove this, as shown in the drawing.

EXPERIMENT 2: Try rubbing one end of a needle on one pole of the magnet, rubbing the other end on the other pole. See whether the needle is magnetized more strongly this way.

EXPERIMENT 3: Try a long needle and a short one, and a thin one and a fat one. Which can be magnetized more strongly?

THE OBEDIENT STRAW

NEEDED: Two soda straws, a jar, a fine thread, a rubber comb, a piece of woolen cloth or fur.

EXPERIMENT: Hang a piece of a straw in the jar as shown. Rub the comb briskly on the wool, then hold it near the jar. The straw inside the jar can be made to move about.

REASON: The comb becomes charged with an excess of electrons which are rubbed loose from the wool cloth. By induction, these repel electrons to the far end of the suspended straw. This makes the near end of the straw positive and attracted to the comb.

This and other static electricity experiments cannot be performed when there is much moisture in the air. They are performed best in a warm room in winter, or in an air-conditioned room where the humidity is kept low.

The Obedient Straw

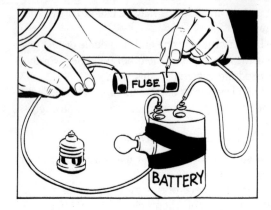

Make a Tester

MAKE A TESTER

NEEDED: A battery and bulb of the same voltage rating, wire, tape, soldering equipment.

EXPERIMENT: Connect the battery and bulb in series as shown in the drawing. If the battery terminals are not screw type the wires should be soldered to the terminals. The ends of the wires can be used to test fuses and some other electrical circuits.

REASON: If the wires are touched together the circuit is completed and the bulb lights. Touch the ends of the wires to the ends of a fuse, and if the fuse is good the bulb lights. If the fuse is burned out the bulb will not light.

Many other objects may be tested with this simple device, and it is safe because the battery voltage is low. Do not try to test anything that is plugged into an electrical outlet. The voltage there, probably 120 volts, is very dangerous.

THE CURIE POINT

NEEDED: Needles, a magnet, a candle or match, pliers.

EXPERIMENT: Magnetize a needle by rubbing one end of it on one end of the magnet. See if it will pick up other needles. Heat the needle over a flame, and again see whether it will pick up other needles.

REASON: If a magnet is heated to the "Curie point" the magnetism disappears. This effect was discovered by Pierre Curie in 1895.

The many magnetic domains in the needle are initially arranged at random, and do not make the needle a magnet to any appreciable degree. When the needle is rubbed on the magnet some of the domains line up so their magnetism points in the same direction.

Their magnetic domains add together rather than cancel one another out.

Heat makes the domains return to the random state. The Curie point, which varies with different alloys, is the point at which the magnet loses its magnetism as the domains return to the random state.

The Curie Point

ELECTRICITY FROM A LEMON

NEEDED: A lemon, a common nail, copper wire, an earphone from a transistor radio.

EXPERIMENT 1: Stick the nail and the end of the copper wire into the lemon. Bend the wire so the other scraped end comes close to the nail. Touch the earphone plug to the nail and wire at the same time, and static sounds will be heard in the earphone.

EXPERIMENT 2: Try this with a potato and other vegetables and fruits.

REASON: The lemon juice, which is the electrolyte for this small cell, reacts at a different rate on the iron and copper, causing a small potential or voltage difference to exist between them.

When the earphone plug makes contact with the metals a current flows in the earphone wires, causing the static sounds. A sound is heard in the earphone when contact is made and again when it is broken. A different sound is heard if the plug is made to slide along the wire and nail.

EARTH MAGNETISM

NEEDED: A wire coat hanger, string, a paper clip, a compass.

EXPERIMENT 1: Straighten the coat hanger, and hang it balanced on a string or small thread. This must be done in a room

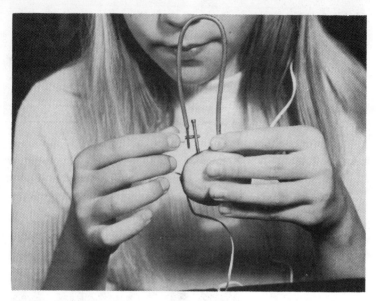

Electricty from a Lemon

where there are no people and no air movements. Overnight the wire will be found pointing approximately north and south like a compass.

REASON: The wire, while not magnetized, still tends to line up with the earth's magnetic lines of force. It is soft iron, and will not remain permanently magnetized, as steel would.

Earth Magnetism

EXPERIMENT 2: Move slowly, so as not to produce air currents. Hold a paper clip or other iron object near the end of the wire, and the wire will be attracted to it.

EXPERIMENT 3: Use the compass to test large iron or steel objects in the house, such as a bath tub or refrigerator. They will be found to be magnetized simply by sitting in one position in the house. Their magnetism comes from the earth's magnetic lines of force.

AN EASY BATTERY TESTER

NEEDED: An "alligator" clip, piece of wire, a flashlight bulb.

EXPERIMENT: Attach a wire to the clip, place the bulb in the clip, and you have a battery tester. Hold the bare end of the wire against the bottom of the battery, touch the tip of the bulb to the metal at the tip of the battery, and, if there is current, the bulb will light.

COMMENT: Do not try this on batteries of higher voltage than the bulb rating. This is suggested for simple flashlight cells, using a regular 1 1/2-volt or 1.3-volt bulb. To test nine-volt batteries a nine-volt bulb must be used, but it is easy to change bulbs.

There is no danger; touch the metal parts of batteries freely. There is no shock.

There are batteries that produce voltages as high as 90 or even 300. There can be shock from them; a 300-volt battery can be dangerous. These are not found in homes or schools, however.

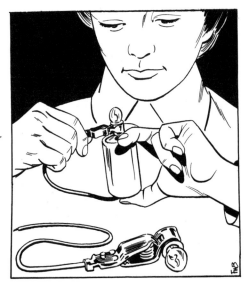

An Easy Battery Tester

A Homemade Compass

A HOMEMADE COMPASS

NEEDED: A sewing needle, a magnet, felt, styrofoam, a dish of water.

EXPERIMENT 1: Rub one end of the needle against one end of the magnet a few times.

This will magnetize the needle.

Break off a piece of styrofoam the size of a pea, stick the needle through it, and float it on the surface of the water. It will move until the needle points north and south.

EXPERIMENT 2: Cut a piece of felt half an inch wide and as long as the needle. Place the needle on it, and it will float on the water for days if the water is not disturbed. This is another way to make the compass on water.

REASON: A magnetized piece of steel will tend to align itself with the lines of force from any magnet. In this case the magnetic lines of force are those of the earth, which run approximately north and south, and line up the needle with them.

A needle without styrofoam or felt may float on the surface of water, held up somewhat precariously by the surface tension of the water. The felt and styrofoam do not wet easily, and as long as they are not wet they float on the water.

THE ELECTRIC GIRL

NEEDED: A girl with long hair, a board, pint Mason jars, orlon acrylic garment, a neon lamp (Calectro NE-51 will do), two more people.

EXPERIMENT: Make a platform with the board and glass jars. Have the long-haired girl stand on it. Have someone rub her hair downward to the waist with the acrylic. The third person may then draw a spark from the insulated girl's finger.

REASON: Rubbing the hair puts a charge of electricity over the girl on the platform. The charge (or part of it) will jump as a spark to someone else. This works only when humidity is very low, usually only in cold winter in a heated room. But it is possible to get 10,000 harmless volts this way.

If the girl holds a neon lamp, and someone else takes the spark from the tip of the lamp, it will flash over and over, each time the spark jumps.

The jars must be clean and dry. The hair and the acrylic must be dry, and the rubbing must be vigorously done. Note how the hair tends to stand out; each hair takes on the same charge, and like charges repel.

It may be more convenient to hold the lamp bulb with a wire handle as shown, but this is not necessary. The high voltage involved is perfectly harmless; the electric current is very low. The effect of the spark may be felt slightly on different parts of the body, however.

The Electric Girl

WHAT CONDUCTS MAGNETISM?

NEEDED: Magnet wire wound around a cardboard tube, battery, compass, objects to test.

EXPERIMENT: Place the tube near the compass, at right angles to the needle. Touch the wires to the battery, and the needle is deflected. Move the tube away until there is only a slight deflection.

Now try inserting various objects into the tube, and see whether the deflection is stronger. Try copper wires or rods, wood, nails, aluminum. The nails, which are iron, will be found to conduct the magnetism, and the needle deflection should be quite strong.

REASON: There is an electromagnetic field in and around the tube when electric current flows in the wire. Materials other than iron are not affected by the field, nor do they affect it. Iron materials themselves become magnetized in the field and make the total magnetic field stronger.

A hard steel object, such as a file, may remain magnetized after the current is turned off.

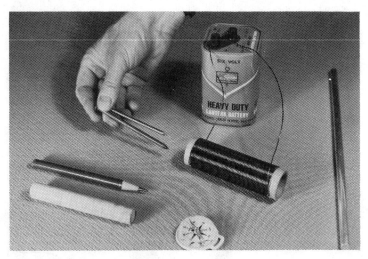

What Conducts Magnetism?

WIGGLE WIRE

NEEDED: A magnet (permanent type), toy train transformer, wire, screws, a wood base.

EXPERIMENT: Mount the magnet on the base, and stretch a wire (not tightly) between screws, a half inch from the pole of the magnet as shown.

Connect one end of the wire to the transformer. As the other end of the wire is touched to the transformer the wire will be seen to vibrate.

REASON: Current carried by the wire sets up a magnetic field around it. It is attracted or repelled by the fields from the permanent magnet according to the rule "like poles repel and unlike poles attract."

Since the alternating current from the transformer (don't use DC) reverses its direction 120 times a second, the wire is alternately attracted and repelled as the magnetic field around the wire is rapidly reversed.

Either a U-shaped magnet or a bar magnet may be used. The wire used by the author in his model was 20 gauge copper magnet wire.

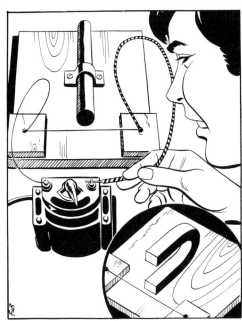

Wiggle Wire

CRAZY PENDULUM

NEEDED: Two bar magnets and string.

EXPERIMENT: Suspend a magnet on a string as shown, and hold it over the other magnet. If they attract, reverse the lower magnet. Then bring them close, and the motion of the upper magnet will be erratic and "crazy."

REASON: The similar magnetic fields repel, and the repulsion is greater as the magnets are close. They move apart, and the repulsion is less, so they come close again.

Crazy Pendulum

The string allows the upper magnet to move in many directions.

There is a device marketed commercially which operates on this principle. One magnet is in the base, and the other in a sphere suspended above it. It is sold as an interesting conversation piece.

The animals shown in the illustration have bar magnets attached to their feet. They are dime store magnets.

A PENCIL MAGNET

NEEDED: Insulated copper wire, a large nail, a pencil, paper clips, dry cells or a toy train transformer.

EXPERIMENT 1: Wind 50 turns of wire around the nail, hold it over the pile of paper clips, connect the ends of the wire to the power source, and a magnet is formed that will pick up the clips, if they are made of iron.

A Pencil Magnet

EXPERIMENT 2: Wind fifty turns of the wire around the pencil and try to pick up the clips. Again we have an electromagnet, but a much weaker one.

COMMENT: A popular school science book states that a magnet cannot be made using the pencil. This experiment shows that it can. The difference is that the iron nail concentrates the lines of force from the wire and focuses them to the end of the winding.

A LOOP MAGNET

NEEDED: Iron filings or bits of steel wool, small gauge copper wire, a six-volt lantern battery.

EXPERIMENT: Bend the wire as shown, and connect one end to the battery. Touch it to the filings, and nothing happens. Hold the other end of the wire to the battery, touch the wire to the filings, and some of them are picked up as the wire is lifted.

REASON: There is a magnetic field around any wire carrying a current. Since iron and steel are attracted by a magnet, they are lifted, because we have made a magnet.

In many books of experiments it is suggested that only a straight wire can be used for this. It may be a difficult experiment with only a wire. But if one turn is put into the wire, as shown, it is easy.

Do not hold the wire against the battery terminal more than a few seconds. It would run the battery down rather quickly if held for very long.

The photo shows a toy transformer used instead of a battery. It is better because it is stronger. But the wire may get hot, so wear a glove on the hand holding the wire.

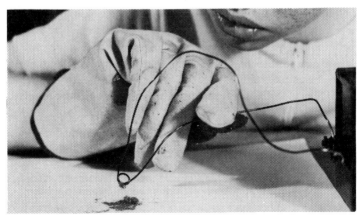

A Loop Magnet

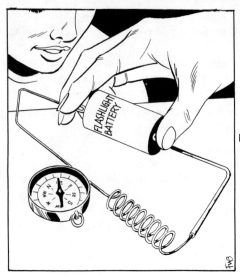

Direction of Magnetic Fields

DIRECTION OF MAGNETIC FIELDS

NEEDED: Battery, wire, a compass.

EXPERIMENT 1: Wind some of the wire into a spring shape. Place the compass close (or better still, under the straight wire) then touch the ends of the wire to the battery. The needle will be deflected.

Place the compass beside or under the spring part of the wire, touch the ends to the battery again, and the needle will be deflected in the opposite direction.

EXPERIMENT 2: Try this with the compass under the wire, then with the compass over the wire.

REASON: Lines of force are circular around the wire, with the wire at the center of the circle. The compass needle lines itself up with the earth's lines of force, which are north and south, but the lines from the wire are stronger and can cause the needle to turn at right angles to the wire.

One cell of flashlight battery is sufficient for this experiment.

BUBBLE MAGIC

NEEDED: Bubble blowing device, a rubber balloon, wool or fur.

EXPERIMENT 1: Charge the inflated balloon by rubbing it with wool or fur, then have someone make a bubble, release it from the plastic bubble maker, then let it fall again to the plastic. Move it close to the charged balloon, and it will be distorted as shown.

EXPERIMENT 2: Release the bubble into the air, bring the charged balloon near it, and the balloon will attract the bubble, causing it to move toward the balloon.

REASON: The negatively charged balloon causes some negative charges on the bubble to move away to the side farthest from the balloon. This leaves the side of the bubble nearest the balloon with a positive charge. Thus the two attract each other. The thin soap film is easily distorted by the slight pull of the static charge on the balloon.

(Remember that static electricity experiments may not work unless the humidity in the room is low.)

Bubble Magic

A SENSITIVE STATIC DETECTOR

NEEDED: Light cardboard, a needle, scissors, a small bottle, a comb.

EXPERIMENT: Cut out the cardboard as shown, and stick a needle through the middle so the point is just below the edge of the card. Rest the needle point on the bottle top. If it does not balance, snip off small bits of card on the heavy side until it does.

Rub the comb on wool or almost any other kind of cloth to charge it. Hold it near the end of the balanced card, and the attraction or repulsion will make the card revolve.

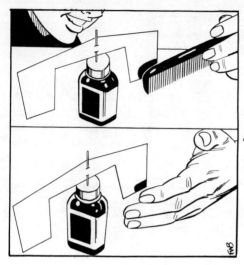

A Sensitive Static Detector

REASON: The rule is: like charges of static electricity repel each other; unlike charges attract. Any charged article will attract any uncharged article such as the cardboard. Since there is practically no friction involved in the turning of this device, very small charges will turn it, even a hand rubbed on clothing on a dry day.

BOARD MAGIC

NEEDED: A light-weight six-foot or eight-foot board, 1 × 3 or 2 × 4 will do, a comb, something to rub the comb with to produce a charge. Glass rubbed with silk or a rubber or plastic comb rubbed with fur or through the hair will do.

EXPERIMENT: Balance the board on a string. Dr. John A. Davis, of Kansas State University, balances the board on a watch glass. Bring the charged glass rod or comb near the end of the board without allowing it to touch. Attraction between the comb and board will cause the board to turn slowly.

REASON: Like charges repel each other; unlike charges attract. The charged comb and neutral board attract because when they are brought close to each other the charged body causes like charges in the originally uncharged body to leave that area, thus leaving the board with an opposite charge. Now the two bodies attract each other.

Static experiments are usually done with small objects. Use of a large pole makes this variation rather amazing. It will work only where humidity is low, and where there is no draft in the air.

Board Magic

MAGNETIC FIELDS

NEEDED: Iron filings, a magnet, and a piece of glass.

EXPERIMENT: Place the piece of glass over the magnet. Drop the iron filings onto the glass and watch what happens. The filings align themselves into a pattern showing the magnetic lines of force (magnetic field) of the magnet.

COMMENT: Iron filings may be obtained from a garage mechanic. Use the tiny iron pieces that fall to the floor when brake drums are turned. Clean them with hot water and detergent and dry them in an oven. To prevent rust, give the filings a last wash in rubbing alcohol to which a few drops of light machine oil has been added.

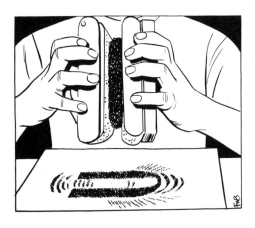

Magnetic Fields

Steel filaments may be used instead of the filings, which can only be used once or twice. These can be obtained by rubbing two steel wool wads together, or by rubbing steel wool between sandpaper blocks, as shown in the drawing.

When handling steel wool, be careful not to breathe the particles or get them stuck in the skin. Never tear a piece of steel wool with the bare fingers; use scissors to cut off what you need.

GLOWING IN THE DARK

NEEDED: A fluorescent lamp in a darkened room.

EXPERIMENT 1: Have the light on. Cover the eyes with a hand to shut out the light. Count off 15 to 25 seconds, switch off the light, and open the eyes quickly. The tube will be seen glowing, the glow gradually but quickly fading away.

REASON: The light given by a fluorescent tube is produced when the electric current flows through a gas containing a small amount of mercury. The light produced is ultra-violet, not suitable as room illumination.

But the ultra-violet light shines on a coating inside the tube, a coating made of phosphors, substances that glow with different colors when excited by the ultra-violet light. The ultra-violet goes

Glowing in the Dark

out as the switch is turned off, but the phosphors continue to glow briefly.

Different phosphors give off different colors, and in the tube, phosphors are used that emit light visible to the eye and suitable for illumination.

EXPERIMENT 2: Try the same experiment with a television set. Its picture is made by phosphors that can be seen to glow briefly after the set is turned off.

ALUMINUM MAGIC

NEEDED: A strong magnet, round pencils, a straight strip of aluminum.

EXPERIMENT: Place the aluminum on the pencils so it can roll easily on them. Hold the magnet just above the strip, and move it quickly back and forth. Do not let it touch the aluminum. Yet the aluminum will move.

REASON: As the magnetic lines of force from the magnet cut through the aluminum, eddy currents of electricity are produced in the strip. An electric current produces magnetic fields. In the aluminum the fields produced interact with the fields of the strong magnet, and pull or push the strip back and forth.

This is the principle of the induction motor. There the currents induced in coils are strong enough to turn the armature of the motor and do work.

An improved way to do this is shown in the drawing. Place a piece of window glass on strips of wood so it rests just above the aluminum. Then rub the magnet back and forth on the glass. This way the magnet does not touch the aluminum yet stays close to it.

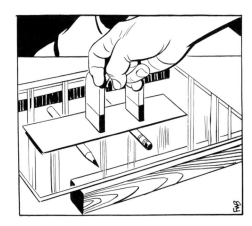

Aluminum Magic

PERMANENT PICTURES OF MAGNETIC FIELDS

NEEDED: A magnet, some white cardboard, iron filings, an atomizer or other water sprayer, salt, a can of aerosol spray paint.

EXPERIMENT: Place the card over the magnet, sprinkle the iron filings on the card, tap lightly and they will assume a pattern representing some of the lines of force of the magnet.

There are two easy ways for preserving the patterns.

1. Spray salt water on the iron filings from the sprayer. Repeat after a few hours and let the experiment stand overnight. Next morning the outline of the filings will remain on the card as rust, and the filings will fall off the paper when the magnet is removed.

2. Set up the card with iron filings as before, but this time spray the card lightly with a dark color paint from the aerosol can. When the paint dries, we have a permanent record of the pattern of the iron filings. Brush off any loose iron filings.

The paint spray method leaves the filings pattern in white on a dark background; the salt water makes rust on white.

ACROBATIC RICE PUFFS

NEEDED: Window glass, a silk cloth, puffed rice, some books, a dry day when the humidity is low.

EXPERIMENT: Place books under the edges of the glass so it rests about 3/4 inch above the table. Spread puffed rice under the glass. Rub the glass briskly with the cloth, and the rice will jump about. Draw a finger along on the glass, and some of the rice puffs will tumble about under the finger.

REASON: The cloth rubs away some of the surface electrons from the glass, leaving it with a positive charge of electricity. Electrons in the puffed rice will be attracted by the positive charge on the glass, and some electrons will move in the rice to the upper surface closer to the glass. The attraction between the positive glass and the negative rice puffs will then be sufficient to make them move about, and jump to the glass. The moving finger will rearrange some of the charge on the glass.

Try putting some of the rice grains in a glass pie dish. Stretch plastic over the dish, and rub it with a finger or a woolen cloth.

Acrobatic Rice Puffs

AN ELECTRIC PENDULUM

NEEDED: Two tall tin cans, three pieces of paraffin wax, a small thread, a paper clip, a rubber balloon, wool cloth, a dry day when the humidity is low.

EXPERIMENT: Set each can on a paraffin block, suspend the clip between them on the thread from the paraffin block as shown below. Rub the balloon with the cloth, bring it close to one of the cans, and the clip will move back and forth between the cans.

REASON: Some of the electrons from the cloth rub off onto the rubber, giving the rubber a negative charge. When it is brought near the can, electrons in the can are repelled to the far side of the can, giving it a charge.

The neutral clip is attracted to it, takes on a similar charge, is repelled from it, and attracted to the other can. There it gives up its charge to the second can, and is attracted again to the first. The ferry-like motion is repeated.

The paraffin serves only as support and insulation, but is very important because of its insulating property.

An Electric Pendulum

ELECTRICITY FROM HEAT

NEEDED: Two wires of different metals, such as iron and copper, an alcohol or gas flame, an earphone.

EXPERIMENT 1: Connect the wires to the earphone, and rub their ends together in the flame. Sounds will be heard in the earphone, showing that a small electric current is produced when the wires are hot and touch each other.

REASON: Each different metal has its own natural rate for losing electrons. Heat increases the loss rate for one metal more than for the other, the more active metal loosing some electrons from its surface to the less active metal as contact is made. This transfer of electrons is the electric current. A pyrometer uses this principle to indicate high oven temperature.

EXPERIMENT 2: Dip the ends of the wires into salt water and rub them together while in the solution. Here electricity is produced, not by heat but by chemical action.

MAKE A BIG MAGNET

NEEDED: A compass, a crowbar, a hammer.

EXPERIMENT: Hold the crowbar pointing north and south, with the north end pointing slightly downward. Hit it several times with the hammer. It will become magnetized and will pick up paper clips or small nails.

Electricity from Heat

REASON: The earth acts as a huge magnet. When the bar is held in alignment with the earth's magnetic lines, and struck, it becomes magnetized as clumps of molecules called "domains" line up, with the north end of one near the south end of the other. If the bar is allowed to lie in a north-south direction it will become magnetized after several days. Hitting it allows the domains to align themselves more quickly. The compass will indicate a weaker charge of magnetism on the bar.

STATIC ELECTRICITY AND THE TV

NEEDED: A cotton string, a TV set, a day when the humidity is very low.

EXPERIMENT: Let the string hang about an inch in front of the TV screen. Turn the set on. As the voices and pictures appear the string moves, either attracted to or repelled from the picture tube. Sometimes it is repelled and then attracted. The action stops after a few seconds. It can take place again as the set is turned off.

REASON: The television picture is made by a stream of electrons that hit the phosphors inside the face of the picture tube. When the set is first turned on the electric charge builds up on the inside face of the tube—a static charge that can attract or repel a string. Some of the charge is soon drawn away by the positive charge on the second anode inside the tube. This charge can be due to 15,000–25,000 volts used in the tube. It is of course harmless because it is never released from the tube in dangerous amounts.

THE WILLIAM GILBERT EXPERIMENT

NEEDED: A strong bar magnet, small iron wires that may be cut from paper clips (if the magnet is quite strong whole clips may be used).

EXPERIMENT: Place the wires carefully along the magnet. They will stand up at various angles in a regular pattern.

REASON: The compass needle "dips" at various parts of the earth, showing the paths of the magnet lines of the earth. The magnetic field of the bar magnet is similar to that of the earth in this respect, and the wires take positions on the magnet representing those of the compass needle on the earth.

Gilbert fashioned a ball of lodestone, a natural magnet, to represent the earth and placed iron wires over its surface. This would make a good science fair project perhaps, but might be difficult. The bar magnet shows it in a simpler way.

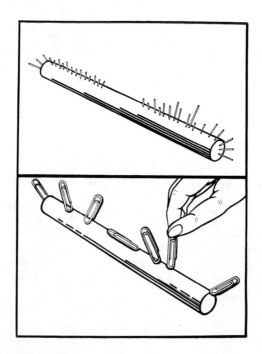

The William Gilbert Experiment

STATIC FUN

NEEDED: Old phonograph record, woolen cloth, a ping-pong ball.

EXPERIMENT: Rub the record briskly with the wool. Hold it near the ball, and the ball will be attracted to it.

REASON: Rubbing the record with the wool gives the record an excess of electrons (negative charges). When it is brought near the uncharged ball, the negative charges on the ball move to the far side, leaving positive charges on the side near the record. The attractive force between the record and ball is sufficient to pull the ball along on the table.

Touch the ball with the record, giving it the same charge as the record, and it should be repelled. The rule is: similar charges repel unlike charges attract.

MAGNET IN THE BATHROOM

NEEDED: A compass.

EXPERIMENT: Hold the compass at different places against the bath tub. The needle will turn, so that it indicates a south pole at some points and a north pole at others.

REASON: The bathtub is a huge piece of steel, and any large piece of steel will become magnetized by the magnetism of the earth. If the tub is in a north-south direction in the room, one end will be north and the other south to the compass.

This applies to the Southern states. In states farther north, where the earth's magnetic lines are more perpendicular, the north and south points of the tub are likely to be top and bottom.

Chapter 9

Air, Air Pressure & Gases

A SODA STRAW ATOMIZER

NEEDED: A glass of water and a soda straw.

EXPERIMENT: Cut the straws in about half. Place one in the water, the other in the mouth, and blow as shown in the drawing. Water will be seen to rise in the straw. If we blow harder, the water will rise to the top of the straw and will be sent in fine droplets across the table.

REASON: Moving air exerts less lateral pressure than the still or less turbulent air around it. The moving air at the upper end of the straw reduces the pressure in the vertical straw, so that atmospheric pressure on the surface of the water in the glass can force some up into the straw.

"Atomizer" is an incorrect word. The water spray is in the form of small drops, not "atoms."

A MILK BOTTLE OCTOPUS

NEEDED: A wet milk bottle, a piece of tissue paper, a match.

EXPERIMENT 1: Twist the paper into a loose rope form and light it with a match. Drop the lighted paper into the bottle, and place the wet hand over the opening. As the flame goes out, the strong suction felt may give some idea of how the octopus' suction cups can draw blood through the skin. The bottle can be lifted as shown in the lower drawing. But don't hold it too long.

REASON: The flame heats the air in the bottle, making it expand so that some of it escapes. When the flame goes out, the cool bottle cools the air, making it contract. This reduces the pressure,

A Soda Straw Atomizer

so that the pressure in, above and around the hand forces the skin downward into the bottle neck.

Don't use too much paper; the flame might burn the hand.

EXPERIMENT 2: Try this experiment without fire, as shown in the photo. Fill the bottle with boiling-hot water, invert the bottle

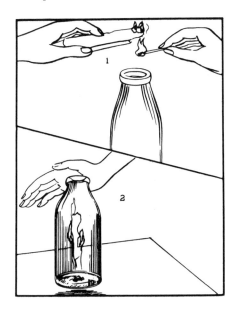

A Milk Bottle Octopus

over the sink so the water gurgles out, then put the wet hand over the mouth of the bottle. Condensing steam and cooling vapor reduce the pressure in this case.

Milk bottles are scarce most places. Other bottles may be used; the author uses a bottle that contained white Karo syrup. Try the Karo bottle in other experiments where a milk bottle is required.

A BALLOON TRICK

NEEDED: A jug or bottle, a balloon, a refrigerator.

EXPERIMENT 1: Cool the jug in the refrigerator an hour. Take it out, put the collapsed balloon tightly over the mouth of the jug, and let it stand. As it warms, the balloon will be blown up.

EXPERIMENT 2: Another way to show the same principle: place the balloon over the mouth of the jug at room temperature, then heat the jug under the hot water faucet.

REASON: As the air in the jug becomes warmer, it expands, and some of it is forced out into the balloon. (Look up Charles' Law.)

BALLOON IN A JUG

NEEDED: A jug, a balloon, hot water, a hard boiled egg.

EXPERIMENT 1: Boil the water, and while boiling hot, pour some into the jug. Shake it around, and pour it out. Place the balloon over the mouth of the jug—quickly. Watch.

Balloon in a Jug

REASON: As the steam and hot vapor in the jug cool, they require less space. Pressure in the jug is reduced, and pressure of the outside atmosphere inflates the balloon—pushing it inside the jug.

EXPERIMENT 2: This is the same principle used to put a hard-boiled egg into a glass milk bottle. Peel the egg first, place it over the mouth of the bottle, and watch it plop in. The egg must be placed over the bottle quickly after the hot water is poured from the bottle.

EXPERIMENT 3: To get the egg out, blow into the bottle while holding it inverted. To get the balloon out of the jug, take it loose from the mouth of the jug and place a pencil down beside it to allow air to flow into the jug around the balloon. The balloon deflates quickly and can be pulled out.

Lung Power

LUNG POWER

NEEDED: Book, rubber balloon, a table.

EXPERIMENT 1: Place the balloon under the book, as shown, and as air is blown into the balloon, the book will rise.

REASON: Air pressure from the lungs is sufficient not only to lift the weight of the book, but also to stretch the rubber of which the balloon is made.

A much heavier weight can be used if a hot water bottle is substituted for the balloon.

EXPERIMENT 2: The author, in one of his school demonstrations in science, places a beach ball between hinged boards and lets a pupil stand or kneel on it. Another pupil can lift the person by blowing breath into the ball.

A PAPER AND COIN TRICK

NEEDED: A coin (a half dollar or quarter will do) and a piece of paper.

EXPERIMENT: Cut the paper so that it is slightly smaller than the coin. Leave the paper flat so that it will lie flat on the coin. Drop both paper and coin separately as shown in the drawing at left, and the paper flutters down more slowly than the coin falls. Drop both together and they fall together, as shown at right.

REASON: The heavy coin, as it falls, takes some air along with it, and the paper rides in this "captive" envelope of air. Sometimes the paper gets separated from the coin during the fall, and then it flutters down slowly because it leaves the air envelope surrounding the coin.

In a vacuum, both would fall at the same rate even though they were dropped separately.

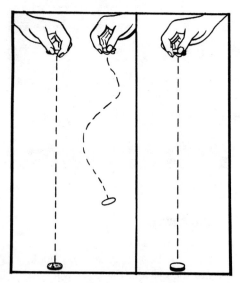

A Paper and Coin Trick

A BOTTLE TRICK

NEEDED: A jar with a tight metal lid, a nail, some water.

EXPERIMENT: Make a nail hole in the lid as shown in the upper drawing. Put water into the jar, and hold it so that the water should run out. It does not.

Make another hole in the lid, as in the lower drawing, and the water pours out.

REASON: A little water may pour out at first, but not much, because the pressure of the atmosphere and surface tension hold it in. If a second hole is made, so that air may get in to make the pressure on the water at the lower hole in the jar exceed that outside the jar, the water can pour out.

And why doesn't the air go in and the water out of the one hole at the same time? The surface tension of the water acts as a film to close the one little hole.

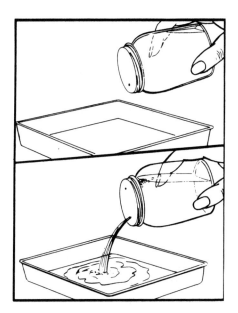

A Bottle Trick

THE CHICKEN FOUNTAIN

NEEDED: A milk bottle, some water, a drinking glass.

EXPERIMENT: Invert the bottle of water in the glass, and a little water will come out. Raise the bottle, and the water in the glass will rise just as far as the mouth of the bottle is raised.

REASON: There is a force on the surface of the water in the glass due to atmospheric pressure. Unless there can be a greater force downward in the neck of the bottle, the water cannot run out. This force cannot exist unless air can get into the bottle, and air cannot easily get in until the bottle is raised above the surface of the water in the glass.

The illustration shows the practical use of this principle in the poultry drinking fountain, where water runs from the jar only as the

The Chicken Fountain

chickens drink from the open part of the container. As the chickens drink from the lower reservoir, air is admitted to the jar, allowing an equal volume of water to run out.

The pressure of air at sea level is equal to a column of water about 34 feet high.

WEIGHT OF AIR

NEEDED: A rubber balloon, a yardstick, a wire, some string.

EXPERIMENT 1: Hang the wire, balloon, and the stick as shown. Move the wire until the stick is balanced. Puncture the balloon. Note that the wire is now heavier than the punctured balloon.

REASON: The air in the balloon was compressed by the balloon; therefore it was more dense and heavier than an equal volume of air at regular atmospheric pressure.

EXPERIMENT 2: Two balloons may be used instead of a balloon and a weight. This experiment is often presented as a demonstration that air has weight. That is not true. It shows only that air compressed by the tight rubber has greater weight than an equal volume of air at regular atmospheric pressure.

INSIDE-OUT BALLOON

NEEDED: A tin can with the end cut out smoothly, a piece of rubber balloon, a nail, some water, some string, a hammer.

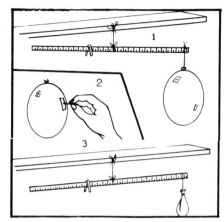

Weight of Air

EXPERIMENT: Punch a hole in the side of the can near the bottom, cover it with plastic tape, and fill the can with water. Stretch the rubber over the end of the can and tie it tightly.

Set the can upright, take off the tape, and as the water comes out of the can, the rubber will be pushed inward by the invisible molecules of air.

REASON: As the water leaves the can, the pressure of the air forces the rubber inward. If there is a hole in the balloon, air will leak in to fill the space left by the receding water, and the balloon will remain straight because the air pressure will then be the same on both sides of it.

If the balloon can be tied on the can as shown in the upper drawing, it will turn inside out as the water leaves the can.

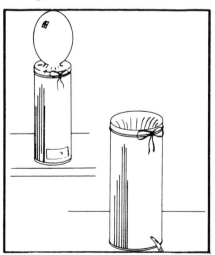

Inside-Out Balloon

The Hanging Glass

THE HANGING GLASS

NEEDED: A glass of water and a flat rubber sink stopper.

EXPERIMENT 1: Wet the stopper, press it down on the glass, and both the stopper and glass may be lifted by the ring.

REASON: When the ring is lifted, the rubber pulls upward slightly, decreasing the pressure of the air in the glass. Atmospheric pressure then presses the rubber to the glass so firmly that the glass of water can lift before the rubber will pull loose.

Wetting the rubber before the experiment makes a better air-tight seal between the rubber and the rim of the glass. Surface tension of the water helps in the lifting operation.

EXPERIMENT 2: If a larger demonstration of this principle is wanted, a rubber suction cup known as a "plumber's helper" may be used. These come with wooden handles. Moisten one, push it down on a smooth surface, and it may be impossible for the normal person to pull it loose. If the area touching the surface is 20 square inches, the pressure of the atmosphere on it can be 280 pounds or more.

A MYSTERY BOTTLE

NEEDED: A soft drink bottle, screen wire, water, toothpicks.

EXPERIMENT: Cover the mouth of the bottle with the screen wire. Fill it with water, as shown in 1. The water pours in freely, and pours out just as readily if the bottle is held as in 2.

But turn the bottle upside down, as in 3, and the water does not pour out. Toothpicks may be inserted through the mesh of the screen; they will float to the top of the water and still the water does not pour out.

REASON: Surface tension of the water is like a thinly stretched rubber sheet. The screen wire increases the surface tension so that the film does not break, and the water does not run out.

The atmosphere exerts small upward force on the lower surface of the film because of the slight decrease in pressure of the air in the bottle.

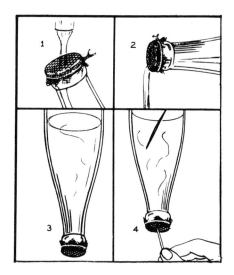

A Mystery Bottle

ELASTIC SOAP BUBBLES

NEEDED: A thread spool, a candle, bubble solution.

EXPERIMENT: Blow a bubble on the end of the spool, hold the spool as shown, and as the bubble gets smaller, the air from it will blow the flame.

REASON: The soap and water film that makes up the bubble is elastic, much as a rubber balloon would be. This is another experiment in surface tension, demonstrating that force due to surface tension draws the surface of a liquid into a spherical shape.

MAKE A DRAFT

NEEDED: A cardboard box, cellophane, cellophane tape, two candles, a match.

Elastic Soap Bubbles

EXPERIMENT: Cut a hole in the side of the box and cover it with cellophane, holding the cellophane with the tape. Make a hole the size of a half dollar in the end of the box. (Make another hole the same size in the other side of the box as shown.) Light both candles. Place one on the table, and place the box over it so that the hole is over the flame. Hold the other candle flame near the other hole, and a draft will be seen which will suck the second flame into the box.

OBSERVATION: As air is heated it expands, becomes lighter, and rises as heavier, cooler air is drawn in to replace it. This is very much the way winds are formed in the air over the earth.

Make a Draft

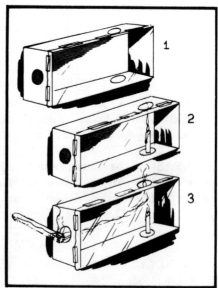

THE RISING PAPER

NEEDED: A strip of paper.

EXPERIMENT: Hold the paper at the mouth, blow over it, and it will rise to the horizontal position.

REASON: Bernoulli learned that a moving air current has less side pressure than the still or slower-moving air beside it. When air is blown above the paper, it has less pressure than the air below. The paper is then pushed up by the greater air pressure below.

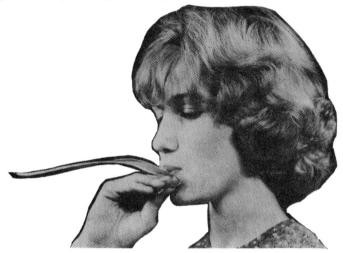

The Rising Paper

WHY NO BLOW-UP?

NEEDED: A sheet of paper and two books.

EXPERIMENT: Place the paper on the books, blow straight under it, and it will bend downward, not upward as expected.

REASON: Bernoulli discovered that air in motion exerts less lateral or side pressure than air at rest or moving more slowly. When air is blown under the paper, it therefore exerts less pressure than the still air above it. The still air then pushes the paper down. This is the principle by which airplanes fly.

THE DANCING BALLOON

NEEDED: A rubber balloon and a warm air register in the floor.

EXPERIMENT: Place the balloon in the stream of rising air, and it will remain there. If the air stream is strong, the balloon will dance up and down in the air.

Why No Blow-Up

REASON: Bernoulli discovered that air in motion exerts less side pressure than still or more slowly-moving air. Therefore, if the balloon moves to one side of the warm air stream, the greater pressure from the slowly-moving air surrounding the moving stream pushes the balloon back.

There is also an irregular up and down dance representing a contest between the downward pull of gravity and the upward movements of the irregular balloon due to the ascending air currents.

The Dancing Balloon

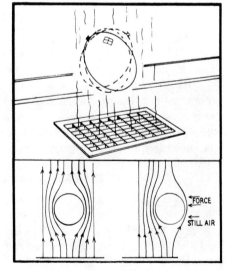

THE MISCHIEVOUS BALL

NEEDED: A funnel, a table tennis ball, a hose-type vacuum cleaner.

EXPERIMENT: Place the hose in the cleaner so that the air blows out. Place the funnel and ball as shown in Drawing A, and the ball will float above it. Place the ball down into the funnel, and it sticks, as in B. Hold the funnel and ball downward as in C, and the ball will be sucked into the funnel, although the air is blowing out.

REASON: The Bernoulli effect is that the lateral pressure of a stream of fluid (air is a fluid) is decreased as the rate of flow is increased. In A, the moving air exerts less pressure than the still air around it, and so the still air pushes the ball back on the moving stream when it tends to move out of it.

B. The moving air around the ball exerts less pressure than the still air pushing down on the top of the ball.

C. The same applies here, except the still air pushes on the bottom of the ball.

The Mischievous Ball

A Heat Motor

A HEAT MOTOR

NEEDED: A paper square, scissors, spool, pencil, needle, thimble, a match.

EXPERIMENT: Cut the paper into a spiral as shown in the upper right drawing. Make a pinhole in the center. Mount it on the pencil as shown; hold a lighted match below it, and the spiral will turn.

REASON: The match warms the air above it. The lighter warm air rises and hits the bottom surface of the spiral, causing it to turn due to the unbalance of forces.

Blowing on the spiral will also make it turn.

THE BLOW-BOTTLE

NEEDED: A bottle and a lighted candle.

EXPERIMENT: Place the tumb over the end of the bottle. Put both thumb and bottle into the mouth. Release the thumb slightly while blowing hard into the bottle, then cover immediately with the thumb.

Bring the mouth of the bottle, with your finger still covering it, close to a lighted candle. Then remove your finger.

RESULT: The air inside the bottle will rush out with enough force to make the candle flicker as though you had blown on it.

While the lungs are not a very good air compressor, they are strong enough to compress enough air for this experiment. While the air in an automobile tire is perhaps 30 pounds per square inch, above atmospheric pressure, the human lungs can offer pressure of only about 1.5 pounds per square inch above atmospheric pressure. (The term "gauge pressure" may be used rather than "above atmospheric pressure.")

THE PARACHUTE

NEEDED: A handkerchief, some string, a weight.

EXPERIMENT: Make the parachute as shown, throw it into the air so that the weight *pushes* the cloth upward. As it comes down, the cloth spreads and the parachute descends slowly.

REASON: As the chute goes up, it is in a compact mass, so that air resistance is slight. When the cloth opens out it must move a larger amount of air out of the way as it comes down, and this means more air resistance and slower motion.

A man coming down with an open parachute hits the ground with about the same speed as though he jumped from a 10-foot height.

The Parachute

THE ELEVATOR CARD

NEEDED: A thread spool, some thread, some cardboard squares, glue.

EXPERIMENT 1: Paste one cardboard square to the spool as shown in the drawing at center right. There must be a hole in the

253

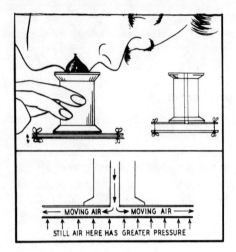

The Elevator Card

center of the card to match the hole in the spool. Hang the other card to the first with thread passed through holes at the corners.

Blow downward through the spool, and the lower card will rise up against the upper card.

REASON: The lower drawing explains; air in motion between the cards has less lateral pressure than the still air that pushes up against the bottom of the lower card. This is the Bernoulli principle.

EXPERIMENT 2: The author presented the simpler version of this in one of his earlier books. Take a 2-inch square of light cardboard, push a straight pin into the center of it, place it on a spool so the pin keeps the card from sliding off, then try to blow the card off by blowing up through the hole in the spool.

A BALLOON TRICK

NEEDED: A rubber balloon, a glass jar, water and soap, a pencil.

EXPERIMENT: Inflate the balloon so it is slightly larger than the mouth of the jar. Try to force it into the jar; it is difficult if not impossible.

Then, slide a pencil down beside the balloon, and the balloon may be pushed into the jar.

REASON: The balloon seals the opening of the jar so air cannot escape around it, and as it is pushed, it tends to compress the air in the jar slightly. It cannot be easily pushed against the air pressure.

When the pencil is used, the trapped air can flow out of the jar through openings at the sides of the pencil, so there is no compression. It may be necessary to wet the balloon with soapy water so it will slide into the jar.

A Balloon Trick

THE WAY THE WIND BLOWS

NEEDED: Two rubber balloons and a means of connecting them together.

EXPERIMENT: Blow up both balloons, and connect them with a tube or spool (as in the drawing). The smaller balloon will shrink while the larger balloon gets larger.

COMMENT: This is the way some of the books tell it, but it does not always work out this way. It does work with two soap bubbles joined by a tube so the smaller bubble feeds air into the larger one. This is because of the difference in surface tension, which is greater in the smaller bubble.

There cannot be a balance in the pressure exerted by two bubble films. There usually is such a pressure balance in the balloons because of the strength characteristics of the balloon surfaces.

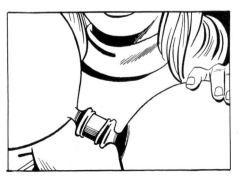

The Way the Wind Blows

Pouring Cold Air

POURING COLD AIR

NEEDED: A cardboard box or styrofoam ice chest, a refrigerator, a warm room or warm day, without wind.

EXPERIMENT: Hold the container at the open refrigerator door for 30 seconds. Place the lid on it carefully, and move it to a warm place where the air is still. Hold it above the face, open the lid, and the cold air will pour down on the face.

REASON: Cool air is a little more dense and heavy than warm air, and so it will pour out of the refrigerator into the container. In the same way it will pour out again, from the container.

If a cardboard box is used it may be closed with a folded newspaper.

THE ARROW

NEEDED: A needle and short thread.

EXPERIMENT: Throw the needle with the point first, at a window curtain. It is likely to go straight as an arrow, and stick into

The Arrow

the curtain. Try throwing it without the thread, and it is not likely to stick into the curtain, but will fall to the floor.

REASON: When the needle is thrown it is given momentum. The thread tends to hold it back, because of air resistance, and this slight pull backward on the eye end of the needle tends to keep the point moving forward. This is the principle of the arrow. Without the thread the needle direction cannot be controlled well. This control by the thread is similar to the balance stick used to guide early rockets.

Wind Resistance

WIND RESISTANCE

NEEDED: Skates and two large cards.

EXPERIMENT 1: Hold the cards as shown and they act as brakes.

EXPERIMENT 2: Hold the cards edge forward, and they will cut through the air easily; not much air has to be moved to allow them to move through it.

EXPERIMENT 3: Hold the cards as shown in the drawing and wave them back and forth. It may be possible to stand up and allow the motion of the cards to move *you* back and forth, an illustration of Newton's third law of motion. (For every action there is an equal and opposite reaction.)

The Gliding Glass

THE GLIDING GLASS

NEEDED: A plastic cup or glass, hot water, a smooth surface such as a plastic counter top or varnished table top.

EXPERIMENT: Rinse the glass with hot water, leave a little water in it, and invert it on the smooth surface. It will "skitter around" as if on ice, with almost no friction. It can be moved with a breath or a feather.

REASON: As the water is poured out of the glass it is replaced by room air. Heat stored in the glass and water heats the air somewhat; it expands, and its pressure lifts the glass a tiny distance from the surface of the table. The glass floats on a film of water and a cushion of air; it may not touch the surface on which it rests at all.

COMMENT: This is the same principle used by "surface-effect-vehicles" or "hovercraft." The English have such a craft to cross the English Channel, from Dover to Calais, replacing the Channel-crossing ships which routinely made most of the passengers sea sick.

Such a hovercraft has a fan which blows air downward into a very large "shroud ring." Although the discharge pressure of the fan is not very high, the total force exerted (pressure times area) is enough to let the hovercraft hover.

A BERNOULLI TRICK

NEEDED: Two empty cans on a smooth table.

EXPERIMENT: Place the cans as shown, blow down between them, and they will tend to roll together.

REASON: Bernoulli discovered that air in motion exerts less pressure than still air. In this case the moving air between the cans reduces the pressure there somewhat, so the greater pressure of the atmosphere on the opposite sides of the can can push against them.

The breath must be blown forcefully to make this trick work. The cans must be close to one another at the start of the experiment.

SAILING

NEEDED: A cart made to roll easily on dowels, a sail mounted on a block of wood resting on the cart, a fan or hair dryer.

EXPERIMENT: Set up the demonstration as shown in the diagram, and by directing the wind as shown, the cart may be made to "tack" into the wind. This means that the cart may be made to travel in a direction which is upwind. This is a principle of sailing.

COMMENT: The writer used two pieces of wood, 1 by 4 inches, with straight dowels between to make the upper board move with very little resistance. The block holding the sail must be movable so the sail can be adjusted to the wind direction.

AIR POLLUTION

NEEDED: A white pan, or any pan with a white paper in the bottom.

EXPERIMENT 1: Fill the pan with water. Set it in the open where it will not be disturbed. Look at it after two or three days.

EXPERIMENT 2: Try two pans, one filled with water and the other dry.

A Bernoulli Trick

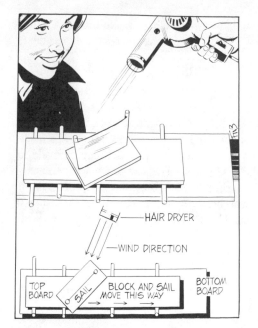

Sailing

Air Pollution

OBSERVATION: Air pollution is often thought of as gases and mixtures of gaseous products, but air always contains solid matter also. It is the solid or "particulate" matter that can be caught and seen in a pan.

In a city the particulate matter from smokestacks can be amazingly heavy. As much as five tons of it can fall on every square mile of a city within 24 hours.

AN ERRATIC BALL

NEEDED: A table tennis ball, a thread or light string, sticky tape, a soda straw with a flexible section.

EXPERIMENT: Attach the ball to the end of the string with a tiny piece of tape. Suspend the ball by the string. Blow upward against the ball, through the straw. Blow slowly, then harder.

REASON: Blow gently up against the ball, just off center, and the ball will try to "outflank" the air jet by going around it to where it can again hang vertically. The ball's motion will be quite erratic.

Blow harder, and the ball will go into the center of the air stream and tend to remain there, even though the stream be moved aside a considerable distance from the normal center of the ball.

In this case the Bernoulli rule will apply: moving air exerts less pressure than still air, and so the ball tends to remain so that the pull of the moving air around it is nearly equal on all sides.

An Erratic Ball

Crush the Jug

CRUSH THE JUG

NEEDED: A plastic jug, a hose attached to it, water.

EXPERIMENT 1: Fill the jug with water, and put a bath spray hose over the mouth. Invert the jug, and as the water flows out through the hose the jug will collapse.

EXPERIMENT 2: Another way to crush the jug with atmospheric pressure: simply suck air out of it with the mouth!

REASON: As water flows out of the jug, pressure above the water level is lessened, so that the greater pressure of the atmosphere crushes the jug.

This is an easy variation of the favorite laboratory demonstration of "crushing the can" with atmospheric pressure. Using the plastic jug instead of a tin can is easier, and the jug may be blown back into shape and used over and over.

If a bath spray hose is not available, the hose may be sealed into the cap of the jug with wax, or may be attached to a metal tube which may be sealed into the cap with wax.

BOUNCE A BALL

NEEDED: A ball such as a basketball or beach ball.

EXPERIMENT: Bounce the ball when it is inflated tightly. Let out most of the air, so the ball is soft, and bounce it again. It will bounce very little.

REASON: As the ball strikes the floor air molecules exert their pressure against the floor through the fabric of the ball. The tightly compressed air in the ball has many times more air molecules exerting this pressure.

Each molecule is in rapid motion and exerts its own pressure. The more molecules the more pressure is exerted. A tremendous pressure is exerted on the body by the moving molecules of air, but this is not felt because there is equal pressure from inside the body.

Air is elastic, that is, it can be compressed then can resume its former volume. A ball bounces as the air expands again after being compressed by the blow against the floor.

ELASTIC AIR

NEEDED: A soft drink bottle with a screw cap, a large nail, hammer, water.

EXPERIMENT: Drive the nail through the cap to make a hole (while the cap is not on the bottle). Put the cap on. Fill the bottle half full of water, turn it upside down, and the water will not pour out. Hold the bottle upside down and blow into it. Take the mouth away, and water squirts out.

REASON: Surface tension across the hole acts like a sheet of rubber to keep the air out of the bottle and the water in. But when air

Elastic Air

is blown into the bottle through the hole, it compresses the air in the bottle above the liquid. Then when the pressure is removed, the air in the bottle expands to force out some of the water.

A fluid (air is a fluid) which tends to assume its former shape (or volume) when squeezed, twisted, or otherwise deformed is elastic.

The person blowing into the bottle may get a little wet. Don't ruin someone's clothes with this trick.

A STRONG LITTLE FINGER

NEEDED: A plumber's suction cup, a smooth surface such as a plastic table top, two people.

EXPERIMENT: Make a small hole in the suction cup. Moisten the cup and the surface on which it is to be placed. Have a strong boy press the cup down on the smooth surface, and pull it away. This can be done easily.

When a small child holds a finger over the hole on the next try, the boy will find it difficult if not impossible to pull the cup from the surface.

REASON: If air cannot get in through the hole or around the edge of the cup, the rubber will stick very tightly to the smooth surface. The atmosphere pushes down at about 14.7 pounds per square inch, and since the mouth of the force cup could enclose an

A Strong Little Finger

area of 16 square inches, the force required to pull it loose from the smooth surface could be more than 200 pounds. This represents the pressure of the atmosphere on the top of the rubber cup.

EASILY CRUSHED CAN

NEEDED: A gallon-size plastic jug with a screw-on lid, boiling water.

EXPERIMENT: Put boiling water into the jug (do not try to boil water in the jug) and shake it with the lid *loose*. When steam and water have stopped coming out, screw the lid on tightly. The jug will begin to collapse. The action can be speeded up by running cold water on the jug.

REASON: As the steam in the jug condenses the pressure in the jug diminishes. Atmospheric pressure crushes it.

This has formerly been done with a metal can, which is crushed with equal ease by the pressure of the atmosphere. But by using the plastic jug a can is not ruined each time. To restore the jug to its former shape, put boiling water into it, screw the lid on tightly, then shake. Or, blow into it.

AIR EXPANSION

NEEDED: A fruit jar with tightly-fitting lid, a balloon, a pan of water, a stove.

EXPERIMENT: Place a few tablespoonfuls of water in the jar, place the jar in the pan of water, and bring it to a boil. Put some air into the balloon, tie it tightly, drop it into the jar, tighten the lid on the jar, and let the jar cool quickly. The balloon will expand and perhaps fill the jar.

REASON: The steam had driven most of the air from the jar, and when it cooled and changed back into water by condensation, it occupied very little space; therefore the air in the balloon expanded to fill some of the space. A gas (air) will fill any space in which it is confined.

AIR PRESSURE TRICK

NEEDED: A milk jug, a plastic bag, a coin, some string, some rubber bands.

EXPERIMENT: Tie the coin on the bottom of the bag as shown, so the string will not pull off of the bag. Place the bag in the jar, blow into the bag to inflate it, place rubber bands around the neck of the jug to fasten the bag, then lift the string. The entire jug will be lifted; the plastic cannot be pulled out of the jug without tearing.

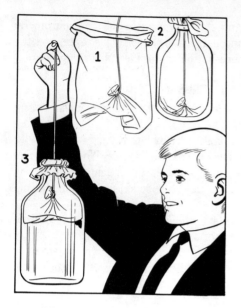

Air Pressure Trick

REASON: As the bottom of the bag is pulled upward by the string, the air pressure between it and the bottom of the jug is reduced. The greater pressure of the atmosphere pressing down on the bag holds the bag in the jug, and the jug may be lifted.

AN AIR CAR

NEEDED: An old long-playing record, a wooden thread spool, a candle, a large rubber balloon, a smooth surface.

EXPERIMENT: Whittle one end of the spool down so that the balloon can be slipped over it. Attach the other end of the spool to the center of the record with candle wax. The holes in the spool and the record should match.

Inflate the balloon, slip its mouth over the spool, place the record on the smooth surface, release the balloon, and the record will glide with very little friction over the surface.

REASON: When the record rests on the surface it tends to remain there because of the friction created when the surfaces move against each other. The air stream from the balloon puts a thin layer of air between the surfaces, eliminating most of the friction.

THE STUBBORN BALLOON

NEEDED: Two cans, a rubber balloon, soap suds, nail and hammer.

An Air Car

EXPERIMENT: Make a nail hole in one can. Put air into the balloon, and try to push it into the can without the hole. It may be impossible. Now try pushing it into the can with the hole. Place the balloon in the end of the can with the hole, suck air out of the hole, and see if the balloon can be drawn into the can that way. It is easier if the balloon is covered with suds.

REASON: To push the balloon into the can, friction between the metal and rubber must be overcome. In the can without the hole, air trapped in the can must be compressed also. As the balloon is pushed in farther the air pressure increases.

If air is removed from the can by suction the balloon may be drawn in, pushed actually by the greater pressure of the air outside the can.

FOUNTAIN IN A JUG

NEEDED: Gallon jug, coffee can, soda straw, a candle, hot water, ice water.

EXPERIMENT: Make a hole in the jug cap, insert the straw, and pour melted candle wax around the straw to seal it in the cap. Put hot water into the jug several times, to heat it thoroughly. Pour out the water, screw the cap on tightly, put the jug upside down in the coffee can of ice water, and watch a fountain flow from the upper end of the straw.

REASON: As the air in the hot jug cools, its pressure is reduced, so that atmospheric pressure on the surface of the ice water can force some of it up the straw. As it comes out of the end of the straw, it cools the air in the jug more quickly, and the pressure is

Fountain in a Jug

further reduced, so that water flows out faster. (Look up Charles' Law.)

If there is steam in the jug some of it is condensed by the cold water, and this adds to the vacuum effect.

CARBONATED BEVERAGE

NEEDED: Two tall glasses (with a fresh soft drink in one and plain water in the other), a small cage made of screen wire, some insects, matches, covers (cardboard is good) for the glasses.

EXPERIMENT: Lower the cage of insects into the glass above the water; the insects are not affected. Lower the cage into the soft drink glass, keeping it above the surface of the liquid, and the

The Carbonated Beverage

insects will soon die. A lighted match will continue to burn above the plain water, but will be extinguished if lowered above the soft drink.

REASON: "Carbonation" of a beverage means that it contains dissolved carbon dioxide gas. Some of the gas constantly escapes, filling the space above the liquid, pushing out the air. The insects die, or the match goes out, for lack of oxygen in the air.

This does not mean that carbonation of a beverage makes it unhealthful. What we drink goes to the stomach; what we breathe goes to the lungs. Our lungs, not our stomachs, require oxygen.

THE BALLOON FALLACY

NEEDED: A yardstick, balloon, paper cup, string, salt, a pin.

EXPERIMENT: Suspend the stick at the middle. Tie a cup at one end and a blown-up balloon at the other. Pour salt into the cup until the stick is balanced.

Let the motion stop, then stick the pin into the balloon. The cup with salt is now heavier. (If pieces of balloon fall to the floor, put them back on the string so the loss of weight cannot be attributed to loss of rubber.)

REASON: This experiment is often given to prove "air has weight." The air in the balloon has weight, of course, but the air outside the balloon has weight, too, and tends to buoy it up. If the pressure inside the balloon could be exactly that of the air outside it, there would be no difference after the ballon is punctured.

What this does prove is that *compressed* air inside the inflated balloon is heavier than the air outside the balloon, which has less pressure.

Another way to try this is to balance two balloons on the stick, and puncture one of them.

ANOTHER MYTH EXPLODED

NEEDED: A bowl, a jar, a lighted candle, water.

EXPERIMENT: Drop wax on the bottom of the bowl to attach the candle. Pour water around the candle, light it, place the jar over it. Air bubbles out at the bottom of the jar, as the air in the jar is heated and expanded by the hot flame. Soon most of the oxygen in the jar is used up, and the candle goes out. Then, as the air in the jar cools and contracts, water is forced up into the jar by pressure of the atmosphere on the surface of the water.

REASON: This proves the air is 1/5 oxygen, since water comes up to fill about 1/5 of the jar. *This is not the correct explanation.* The water comes up into the jar mainly because of the expansion and contraction of the air in the jar.

Another Myth Exploded

However, the oxygen is mostly consumed by the flame, and about an equal volume of carbon dioxide is produced. Carbon dioxide is about 26 times as soluble in water as the oxygen, and so, some of the carbon dioxide is dissolved in the water, adding somewhat to the lessening of the air pressure in the jar.

A BAD BAROMETER

NEEDED: A quart milk bottle, a balloon, soda straw, a box, a wax candle, rubber bands to fasten the balloon.

EXPERIMENT: Fasten a section of balloon rubber over the mouth of the bottle with the tight rubber bands, and attach the end of the straw to the center of the rubber with melted wax. Mark the

A Bad Barometer

height of the other end of the straw on the box, and as the barometric pressure varies, the straw will move very slowly up and down.

REASON: This is often suggested as a simple school science project. As a barometer it cannot work, because variations of temperature move the straw more than barometric pressure. As temperature rises, air in the bottle expands, pushing the rubber diaphragm up and the pointer down. No commercial barometer is made exactly like this.

THE ERRATIC BALLOON

NEEDED: A helium-filled balloon and an automobile.

EXPERIMENT: Hold the balloon by the string in a closed automobile. Let someone drive the vehicle in a circle. All objects in the car will tend to move toward the outside of the circle—except the balloon. It will float in the opposite direction.

REASON: The balloon tends to rise because its helium is lighter than air. As the automobile turns, the heavier air tends to move toward the outside of the circle, as do the people and other objects. The helium, being lighter, tends to move in the opposite direction.

AN ANALOGY: Float a ping-pong ball on water in a fish bowl, and turn around with the bowl in a hand. The water will move toward the outside of your circle, and the ball will move toward you.

THE GHOST FLAME

NEEDED: A candle or gas flame, piece of screen wire, a pair of pliers, a match.

EXPERIMENT: Hold the screen as in the upper drawing, so that half of the flame is cut off. There is no flame above the screen. Light the place above the screen where there is no flame, and a ghostly flame appears.

REASON: When the screen is held midway of the flame it conducts so much heat away that the flame ends there. The gases do

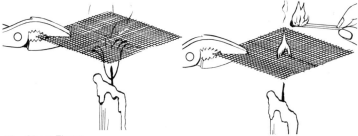

The Ghost Flame

271

Smoke Ring Cannon

not get hot enough to ignite, but they are there, passing through the screen mesh. When lighted they will burn. If a gas flame is used, the gas may be lighted above the screen, and it will not catch fire below it, for the same reason.

SMOKE RING CANNON

NEEDED: A coffee can with two plastic covers, incense or something else to produce smoke.

EXPERIMENT: Cut the bottom of the can out. Place plastic lids over both ends of the can; one lid should have a hole in it. The hole can be the size of a dime if the can is small and the size of a quarter if it is a large one. Put smoke into the can, and rings are produced by tapping the lid opposite the hole.

REASON: When the lid is tapped, an air wave comes out at the hole. The wave is similar to a low frequency sound wave.

A little smoke comes out with each wave. The edges of the hole drag on the edges of the disc of smoke and air, so that a toroidal or "doughnut" circulation is begun. This shape continues as the rings move away slowly and beautifully. See also "Smoke Rings in Water."

Chapter 10

Heat

EXPLOSIONS ON THE KITCHEN STOVE

NEEDED: Popcorn, a covered skillet or popper, a stove.
EXPERIMENT: Pop the corn.
OBSERVATION: Most substances when heated give off their moisture slowly. The water in a grain of popcorn, however, is enclosed in an air-tight sheath which does not explode until a rather powerful steam pressure is built up inside it.

As the sheath bursts, most of the cells inside it explode, too, from the steam pressure inside them. The expanding cells form the delicious white mass.

The pressure that builds up inside the popcorn before it pops has been estimated at from 15 to 100 pounds per square inch.

THE OBLIGING BALLOON

NEEDED: A small rubber balloon, two water glasses, hot and cold water.
EXPERIMENT: Blow up the balloon. Heat both glasses by pouring hot water into them. Place them quickly on either side of the balloon, hold them securely, then cool the glasses under the cold water faucet.

The sides of the balloon will be sucked up into the glasses, and will hold so tightly that one glass may be lifted with the other as shown in the drawing.

REASON: As the glasses are cooled under the cold water, the air in them cools and contracts. The reduced pressure allows the pressure of the atmosphere to force parts of the balloon tightly into the mouths of the glasses.

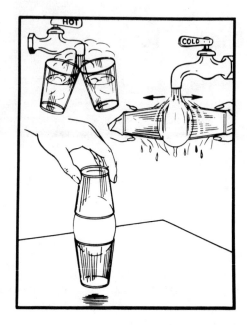

The Obliging Balloon

TWO FRICTIONS

NEEDED: A piece of bare wire. A wire coat hanger will do.

EXPERIMENT 1: Pull the straight wire through the hand. Heat will be felt.

EXPERIMENT 2: Bend the wire back and forth quickly several times. It will get warm.

REASON: The answer is friction in both cases. Friction between the wire and hand generates heat. In the second experiment the friction is between molecules of the wire as they move over each other.

In both instances mechanical energy is converted into heat energy by means of friction.

MYSTERY ICE

NEEDED: Soft drinks in the freezer.

EXPERIMENT 1: Take out a bottle of the drink just before it begins to freeze. Ice will form in it when the cap is removed.

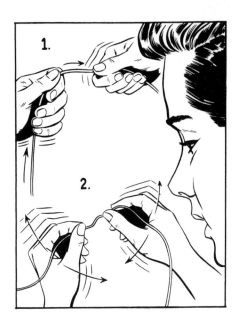

Two Frictions

EXPERIMENT 2: Don't let any liquid freeze solid in a glass container in the home freezer. Water expands when it freezes and will break the strongest glass bottle. To show this, put water into a bottle, wrap the bottle several times with a cloth to protect from broken glass, and let it freeze.

REASON: Carbon dioxide gas dissolved in the drink lowers its freezing point. When the bottle cap is removed, some of the carbon dioxide will escape, and this raises the freezing point so that it can freeze at a higher temperature.

AN OLD FREEZING QUESTION

NEEDED: Boiled water, boiled and aerated water, tap water, salt water; four similar glasses.

EXPERIMENT: Boil some of the water for two minutes. Fill one of the glasses with it. Put some of the boiled water into a jar, tighten the lid, and shake it vigorously to get some air dissolved in it. (This should be done after the water has cooled.) Fill a glass with this aerated water, another with plain tap water, and another with salt water.

After all the glasses of water have been allowed to reach the same room temperature, place them into a freezer.

OBSERVATION: The boiled water may freeze first, the aerated water second, the plain water third, and the salt water last. The reason is that most substances added to water lower its freezing point. This includes air. Boiling releases most of the air that is dissolved in water.

THE FIREPROOF HANDKERCHIEF

NEEDED: A half dollar, a handkerchief, a match or cigarette.

EXPERIMENT: Wrap the coin in a single thickness of the cloth, and draw the cloth tightly around the coin. A match flame held to the tight part of the cloth for a short time will not burn it. The lighted end of a cigarette may be placed against the cloth without burning a hole in it.

REASON: While the cloth does not transmit heat very well, it lets enough go through into the metal coin to keep the temperature of the cloth below its burning point. The metal conducts the heat away from the cloth which is in contact with the coin.

(Suggestion: do not use a valuable handkerchief for this experiment—it *might* get scorched.)

See the heat conduction experiment in the tricks section of this book.

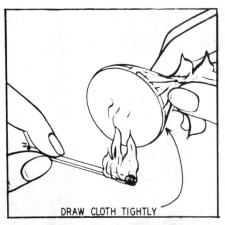

The Fireproof Handkerchief

SQUEEZE BOTTLE

NEEDED: An empty soft drink bottle, a dime, some water.

EXPERIMENT: Place the dime over the mouth of the bottle, and drop a little water around the edge to seal it. Grasp the bottle with both hands, squeeze it, and the dime will dance up and down.

REASON: Squeezing the bottle probably has no noticeable effect on the dime at all, but the warmth of the hands causes the air inside the bottle to expand after the squeezing has been going on for several seconds (Charles' law). The escaping air causes the dime to move up and down.

Charles' law teaches that the volume of any gas is directly proportional to its absolute temperature.

THE HOT WEIGHT

NEEDED: Wire, a weight, ice cubes, a hair dryer.

EXPERIMENT 1: Hang the weight on the wire so it barely swings above a bare table or floor (no carpet). Heat the wire with the hair dryer, and it will expand so that the weight drags the floor or table.

EXPERIMENT 2: The wire may be cooled enough by running the ice cubes over it to make it contract so the weight swings freely again.

REASON: Most substances expand when warmed and contract when cooled. Iron, copper, or aluminum wire will demonstrate this in this experiment.

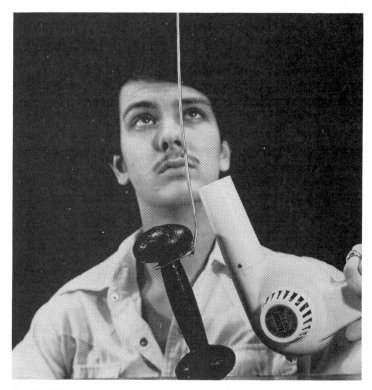

The Hot Weight

Hot or Cold

HOT OR COLD

NEEDED: Breath on a cool day.

EXPERIMENT: Hold the hand in front of the open mouth and blow. The air is warm. Blow through pursed lips, and the air is cool.

REASON: Air coming from the open mouth comes at little less than the temperature of the inside of the body, which is warm. But when air is forcibly blown in a thin steam through the lips it feels cooler for two reasons.

The air is compressed slightly and cools when it expands outside the mouth. This is one of the laws of science: a gas cools as it expands.

The air stream coming rather fast from the lips gathers surrounding air into the stream, and that surrounding air is cool.

FREEZING

NEEDED: Olive oil.

EXPERIMENT: Place the bottle of oil in the refrigerator (not the freezing compartment) and let it freeze. Note that when it is removed and begins to thaw, the frozen oil will be on the bottom, and not floating on the top as in frozen water.

REASON: Most liquids, when they freeze, become solid from bottom to top, because they contract on freezing. Water expands when it freezes, so is lighter, and the solid ice floats on top. This is fortunate for living things in water. If water did not freeze in this way, our lakes and rivers could become solid ice.

Freezing

CRAZY WATER

NEEDED: Hot and cold water spigots.

EXPERIMENT: Turn on a little water from both spigots. The cold water will continue to run, but chances are the hot water will turn itself off after a few seconds.

REASON: The water flow is controlled by a metal-and-rubber plunger that moves in or out with a screw as the handle is turned. Opening the spigot (more correctly called a faucet) moves the plunger out so water can flow past it.

Hot water causes the plunger to expand so that it closes the water opening. The cold water does not cause the cold water plunger to expand.

THE WARM SLEEVE

NEEDED: Two sleeves of different weaves, sunlight.

EXPERIMENT 1: Hold the arms in the sun. The arm covered with tightly woven material will get warm more quickly than the other.

EXPERIMENT 2: Try this with a dark and a light sleeve in the same weave. The dark will absorb light, changing it to heat, much more than the light color.

REASON: The loosely woven material has many air spaces between the threads, and air is an insulated against heat. The solidly woven material conducts heat more readily.

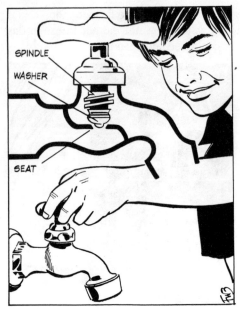

Crazy Water

It is important to use material of the same color for this comparison, since some colors absorb more heat than others and would conduct it on to the arms. A white sweater with tight sleeves, and a cotton shirt sleeve drawn tightly against the arms were used in the author's test of this experiment.

STARCHY BUBBLES

NEEDED: A teaspoon of starch in half a glass of water, a saucepan with handle, an electric stove.

The Warm Sleeve

EXPERIMENT: Let the starch and water mixture come to a boil. Press down on the pan, and the boiling is more rapid as shown by the increase of bubbles. The starch makes the bubbles more visible.

REASON: Pressing down on the pan makes better contact of the pan with the heating element. The elements are seldom perfectly flat, and when the pan is pressed, more of the heating surface touches the pan, for two reasons. Pressing tends to flatten the element, and thus gives more contact. Also, pressing tends to flatten the pan against the element somewhat, although this cannot be seen with the eye.

This second type of flattening takes place when any two substances are pressed together. Even a heavy railroad rail is flattened somewhat when the wheel rolls over it. The metals usually return to their original shape when the pressure is removed.

Starchy Bubbles

FAST COTTON

NEEDED: Strips of polyester cloth, strips of cotton cloth, water.

EXPERIMENT 1: Dip the cloth pieces into water. The cotton soaks up water immediately; the polyester does not.

EXPERIMENT 2: Get a cotton cloth and a permanently pressed cloth (shirt tails of good shirts will do—this will not damage the shirts) and dip them in water. The plain cotton will soak up water more quickly.

REASON: Why does a dress or shirt made of polyester material feel hotter on a summer day than one made of cotton? The structure of the threads in the different materials is quite different. The cotton cloth fibers have spaces that allow water to flow up between them by capillary attraction and heat from the body to flow out from the body through the fibers. The polyester fibers are more solid. The wrinkle-proofing substances in the permanently pressed cloth will slow or prevent the easy flow of water through the fibers by capillary attraction.

Fast Cotton

ICY TEMPERATURE

NEEDED: Paper cup, refrigerator or freezer, thermometer, water.

EXPERIMENT 1: Place the thermometer in the cup of water, let it stay in the freezer overnight. Check the temperature. It is likely to be considerably below the freezing point of water, 32 degrees Fahrenheit or 0 degrees Celsius. The temperature should go to 32 degrees when the ice begins to melt, and should remain at that temperature as long as the thermometer bulb is encased in ice.

EXPERIMENT 2: If an oven thermometer is available an experiment on the temperature of heated water is possible. Heat some—the temperature rises until it is about 100 degrees Celsius (212 degrees Fahrenheit). This varies with atmospheric pressure. The water temperature remains at that level; additional heat applied is used in the boiling.

Water under pressure can be at a much higher temperature, however.

REASON: Ice can be much lower in temperature than its freezing point, but should not go higher than that as long as it remains ice. Additional heat absorbed by it will be expended in the melting process after it reaches 32 degrees.

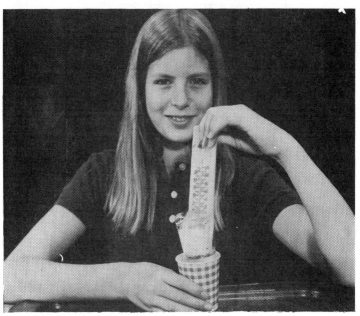

Icy Temperature

DARK HEAT

NEEDED: Two like metal cans, dull black paint, two thermometers, water, sunlight coming through a window.

EXPERIMENT 1: Spray one can black inside and out. Fill both cans with tap water, an equal amount in each. Place the thermometers in the cans, and put them side by side in the sun. Take readings every ten minutes, and it will be found that the water in the black can warms more rapidly.

REASON: Much of the light that falls on the bright metal can is reflected away, while more of the light falling on the black can is absorbed by the black paint and changed into heat.

EXPERIMENT 2: Put warm water in both cans, and set the cans on a table, being careful not to use water hot enough to break the thermometers. The black can will radiate heat away more quickly and will cool faster.

A WATER DANCE

NEEDED: A skillet, heat, water.

EXPERIMENT: Heat the skillet so that when a drop of water falls on it the drop will dance around as it evaporates. See how long it takes the drop to evaporate. Let the skillet cool some, drop water on

Dark Heat

it, and the drop will evaporate faster if it does not dance, even though the skillet is cooler.

REASON: As the drop touches the hottest skillet a little steam is formed, under the drop, enough to hold the bulk of the drop above the hot surface, in effect, insulating it somewhat so heat travels into it comparatively slowly.

If the drop does not dance, it touches the hot metal over a large area of its surface, and so absorbs heat faster, and evaporates faster.

A Water Dance

FREEZE TO FINGER

NEEDED: An ice tray just out of the freezer.

EXPERIMENT: Take the cubes out. Touch the finger to one, and it is likely to freeze to the finger. Try wetting the finger before touching the cube, and see if it sticks more tightly to the finger.

REASON: As the finger approaches the cube the warmth of the finger melts a tiny bit of the cube. As heat is conducted away from the contact through the cold cube the tiny bit of water freezes into ice.

It is possible to have a hand frozen to a very cold piece of metal so that skin is pulled off the hand. The ice cube stunt is safe, however, because the finger continues to get a little warmer at the contact point all the time and will again melt the ice at the contact point.

See the experiment "Melting Under Pressure."

Freeze to Finger

A RUBBER BAND MYSTERY

NEEDED: A weight suspended from a rubber band, a match.

EXPERIMENT: Hold the flame of the match near the rubber band, and as the band is heated, it contracts, lifting the weight.

REASON: While most substances expand when heated, some will contract. Eleven volumes of ice, when heated, make only about ten volumes of water, for example. Such substances as rubber contract because of their increased stiffness when heated; therefore, more force is required to stretch the rubber band. For better technical understanding, look up Young's modulus.

ANOTHER RUBBER BAND TRICK

NEEDED: A thick, wide rubber band or piece of inner tube rubber.

EXPERIMENT: Stretch the rubber band quickly while holding it to the lips. It will get warmer. Let it contract, and it cools.

REASON: Here's one of those questions that cannot be answered satisfactorily and simply. Gases, when compressed, get hot, and cool when allowed to expand. It is thought that the rubber molecules act in the same way, but this is not definite.

If you increase the temperature of a rubber band its tension increases. If you increase the tension of a rubber band (with all else constant) its temperature increases.

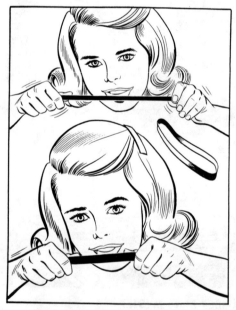

Another Rubber Band Trick

A FIREPROOF CLOTH

NEEDED: Two ounces of rubbing alcohol, one ounce of water, a square of cotton cloth.

EXPERIMENT: Mix the water and alcohol and soak the cloth in it, hold it at a distance with a coathanger and try to light it with a match. It seems to burn, yet the cloth remains unharmed.

REASON: The alcohol burns when lighted with the match, but the heat produced is not sufficient to evaporate the water from the cloth, which keeps the temperature of the cloth so low that it does not burn.

Do not try to fireproof clothing in this manner.

Fireproof Cloth

FIREPROOF PAPER

NEEDED: Two pieces of paper, two matches, a drinking glass.

EXPERIMENT: See how easily the paper burns when lighted with a match. Wrap the other piece around the glass, so that it touches at all points, then try to light it. It will not burn.

REASON: The glass absorbs the heat from the match as it goes through the paper, so that the paper does not reach its kindling temperature. Also, the glass tends to cut off the oxygen of the air necessary for burning.

BURNING PAPER

NEEDED: A tightly rolled paper, a crumpled paper, a match.

EXPERIMENT: Try to light the rolled paper with a match, and it is very difficult. Light the crumpled paper, and it burns readily.

REASON: When heat is applied to the rolled paper it does not burn easily for two reasons. First, the air cannot get to the inside sheets, and they cannot burn without oxygen from the air. Second, when the paper is rolled, much of the heat from the match is conducted into the roll, and the outside layers of paper cannot get hot enough to ignite. The crumpled paper is heated quickly by the match, and since air is all around it, it burns easily.

HOW HOT THE WATER?

NEEDED: Two paper cups, water, salt, a heat source.

EXPERIMENT: Boil water in the paper cup in the regular way. The paper does not burn. Heat dry salt in the other paper cup; the paper burns.

REASON: Water cannot be heated above its boiling point—any more heat applied is used up in boiling the water. Since the burning point of the paper is higher than that of boiling water, and since the paper conducts the heat readily into the water, the paper does not burn. The salt does not conduct the heat away from the paper readily, and has a melting point of about 800 degrees centigrade which is much higher than the burning point of the paper. Therefore, the paper is heated to its burning point.

How Hot the Water?

THE STUCK JAR LID

EXPERIMENT: A jar lid that cannot normally be removed when cold can be made loose enough to be turned when heated under hot water.

REASON: First, the lid expands when heated, as most substances do, and may become looser for this reason. Second, if there is a rubber gasket in the lid, it will contract when heated, making the lid loose enough to be twisted off. Remember rubber is one of the few substances that will contract when heated.

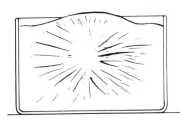

Contract or Expand?

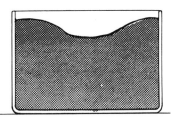

CONTRACT OR EXPAND

NEEDED: A dish of water in the freezing compartment of a refrigerator, a dish of melted paraffin.

EXPERIMENT: Let the water freeze in the refrigerator, and let the paraffin freeze at room temperature. The water expands on freezing, and the paraffin contracts.

REASON: Water, type metal, and cast iron are among the very few substances that expand when they freeze. Most others continue to contract as their temperatures are lowered.

Water contracts until its temperature is lowered to four degrees centigrade, then it begins to expand. While freezing it expands nearly 10 percent, then, as its temperature is further lowered, it contracts again.

The expansion of water on freezing is of great importance to the life forms in lakes and ponds. If it would contract, as most substances do, the lake would freeze solid from the bottom upward, making it impossible for fish to live.

A WARM CUSHION

NEEDED: A kitchen grater, old styrofoam containers, a cloth sack for making the cushion.

EXPERIMENT: Grate the styrofoam into small pieces, stuff them into the cloth, and sit on the cloth. The cushion seems to generate heat.

REASON: The styrofoam is a good insulator for heat, so most of the body heat is retained rather than dissipated. It is a "heat trap."

WHY THE ALCOHOL RUB?

NEEDED: Rubbing alcohol and water.

EXPERIMENT: Extend your hands, and have someone pour water on one and alcohol on the other. If both liquids are the same temperature the alcohol will feel much cooler to the hand.

REASON: Alcohol evaporates faster than water, and when a liquid evaporates it must take heat from its surroundings. Therefore, the alcohol takes more heat from the hand *faster* than does the water.

Nurses rub patients with alcohol because it is cooler, and it also has some germ-killing power.

REFLECTION OF HEAT AND COLD

NEEDED: A heat source such as a corn popper or hotplate, a pitcher of ice cubes, a curved piece of metal or cardboard, some kitchen foil, a book, two thermometers.

EXPERIMENT: Cover the curved card or metal (the author used the side of a coffee can cut with a can opener and tin snips) with crumpled foil to be used as a reflector. Arrange the heat source, the reflector, and a thermometer as shown so that the heat reflecting from the foil will reach the thermometer. A book between the heat source and the thermometer will prevent the direct heat radiation from affecting the thermometer.

The heat will be reflected and the higher temperature on the thermometer will be easily read if the heat source, reflector, and thermometer are in the right positions. Remove the heat source and place the pitcher in its place. Use the other thermometer which has not been heated and see if it will show a drop in temperature from the reflected cold from the ice cubes.

REASON: There was no change in the temperature shown by the second thermometer. Radiant heat is energy and can be reflected. Cold is the relative absence of a certain amount or degree of heat, and cannot radiate anything.

Chapter 11

Light

A SKIN DIVER'S MYSTERY

NEEDED: A glass of water and a pencil.

EXPERIMENT: Place the pencil in the glass and look up at it as shown in the drawing. The pencil—and other objects—cannot be seen when they are above the surface of the water.

REASON: For light striking the under surface of water at an angle of about 48.6 degrees or more, that surface acts as a mirror, reflecting the rays back. The exact angle depends on the temperature of the water.

This is why a skin diver who looks up can see out of the water only at an angle less than 48.6 degrees from the vertical.

A SAUSAGE FROM THE FINGERS

NEEDED: Just two fingers.

EXPERIMENT: Hold the fingers together in front of the face, at arm's length, and look at a distant object, as in the diagram. The fingers seem to form a sausage as in the top drawing. Draw the fingers apart somewhat, as in second diagram, and the sausage becomes a ball. Draw them farther apart, as in third diagram, and the ball disappears.

REASON: The dotted lines in the diagrams show the path of light from the distant object (where the lines converge) to the two eyes. The circles represent the head. This is an optical illusion.

A Skin Diver's Mystery

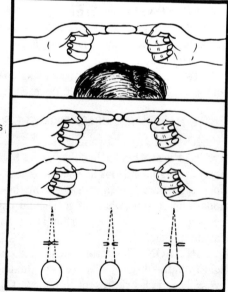

A Sausage from the Fingers

GLASSES THAT MAGNIFY

NEEDED: A round glass container, a square glass, water, a pencil.

EXPERIMENT: Place the pencil in the square glass of water, and it looks the same size. Place it in the round glass of water, and it looks larger where it is under the water.

REASON: The curve of the glass acts as a magnifying glass. The lower drawings show the real and apparent paths of light coming from the pencil to the eye through the glasses.

There is refraction or bending of the light at both surfaces of the glass, changing the direction of the light ray. The light appears to come to the eye along the dashed line, as though the pencil were larger.

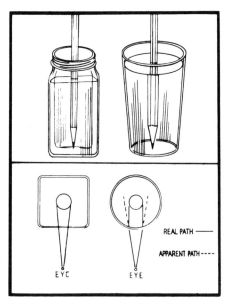

Glasses That Magnify

BIRD IN A CAGE

NEEDED: Twine, a piece of cardboard, ink or crayon.

EXPERIMENT: Draw a bird on one side of the card, and a cage on the other. Attach the strings as shown, twist them, and pull on the ends to make the card spin around rapidly. The bird will be seen in the cage.

REASON: The eye continues to see an object after it has disappeared. This lasts only a fraction of a second, and is called "persistence of vision." Therefore, the eye continues to see both pictures on the card as it turns.

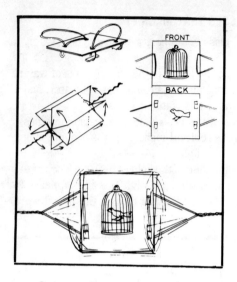

Bird in a Cage

It is persistence of vision that allows us to see movies and television as constant pictures that move, instead of what they are: a series of flickering pictures, alternating with blank screens.

WHY THE DARK SPOTS?

NEEDED: A piece of glass, some drops of water, a light surface, some sunshine.

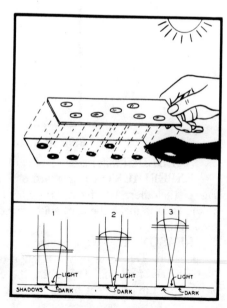

Why the Dark Spots?

EXPERIMENT: Place drops of water on the glass, and hold it above a sheet of white paper or other light surface. The water is clear, yet there will be dark spots on the paper below the water drops, many with an extra light spot in the center. Move the glass up and down.

REASON: The sunlight striking the glass shines through evenly. But the light striking the drops is refracted so that it is concentrated in the center and surrounded by dark circles. The diagram in the lower part of the drawing shows this refraction.

The Fat Reflection

THE FAT REFLECTION

NEEDED: A shiny automobile door.

EXPERIMENT: Look at a reflection of a person in the door of the car. The person will look fat.

REASON: The diagram shows the paths of the light rays which travel from various parts of the body to the car door and back to the eye of the viewer.

Rays from the curved door are spread outward, up and down, but not right and left, because the door surface is not spherical. The width of the reflection is about the same as the person, but the height is much less, making it out of proportion.

295

TWO MIRROR TRICKS

NEEDED: Two mirrors, some cellophane tape, a short pencil.

EXPERIMENT 1: Tape the mirrors together at right angles to each other. Stand them up, and place the pencil between them. How many pencils do you see?

EXPERIMENT 2: Look into the mirrors, and the face will be seen, but not as in an ordinary mirror.

REASON: Light leaving the pencil goes in all directions. Some passes directly to the eye, some passes to a mirror then to the eye, and some may be reflected from both mirrors before reaching the eye.

The face is reflected from both mirrors before reaching the eye. This gives an effect opposite to that from a single mirror. A finger touching the left side of the face seems to touch the left side of the mirrored face.

Two Mirror Tricks

THE CORNER REFLECTOR

NEEDED: Two dime store mirrors, a flashlight.

EXPERIMENT: Arrange the mirrors as shown, so the angle is 90 degrees between them. They may be held in place with tape. Then, when the light is shined at the angle where the mirrors meet, or anywhere within the angle created by the mirrors, the light reflects back to its source. If the light is held below the arrangement, another mirror placed on top of the two is necessary. If the light is held above, another mirror placed below the two is necessary.

An array of such mirror combinations was left on the moon so a laser beam could be reflected back to earth from it.

REASON: A more technical explanation is: the corner reflector has the property—from elementary geometrical optics—that a ray of light impinging on it is reflected back parallel to the incident ray.

Small mirrors are good for this experiment.

THE DOUBLE IMAGE

NEEDED: A piece of thick, flat glass, black paper, flashlight.

EXPERIMENT: Place the black paper behind the sheet of glass and you have a mirror. But shine the flashlight on your new mirror as shown in Drawing B. You will see that two images are reflected from the mirror.

REASON: Both the top and bottom surfaces of the glass act as mirrors. The light rays coming from the flashlight are reflected and refracted as shown in Drawing C.

Light going from a transparent substance into another transparent substance of a different density is bent or refracted. This may be seen in Drawing C.

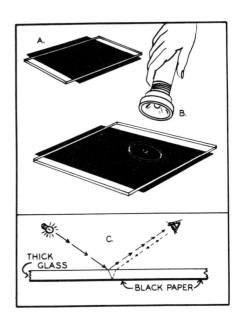

The Double Image

A ROCK REFLECTOR

NEEDED: A small stone or other rough object, clear nail polish.

EXPERIMENT: Paint half the stone with the polish, or simply wet half of the stone with plain water. Note how much more glossy the painted or wet half appears.

REASON: Light striking the rough stone is reflected in an irregular fashion which is called diffused reflection. This is how we see most objects. The polish or water on the stone makes the surface appear as many small mirrors reflecting the light in a more regular pattern as from many small smooth surfaces.

A Rock Reflector

AN ILLUSION

NEEDED: Smoked glass or dark sunglasses, a weight on a string (the author used a bar of soap wrapped in foil).

EXPERIMENT: Have someone cover one eye with the dark glass, then, leaving both eyes open, watch the weight. Swing the weight in a straight arc perpendicular to the other person.

He is likely to see the weight swinging in a circle. Have him switch the glass to the other eye, and the weight will seem to go around in the other direction.

COMMENT: The author has not found agreement among scientists as to the cause of this illusion. He invites opinions. Note that not everyone will observe the circular motion of the weight.

This works better if the people performing it are several feet apart. The pendulum must swing at right angles to the line of sight, not to and from the viewer.

An Illusion

THE TYNDALL EFFECT

NEEDED: Two coffee cans, tape, an electric lamp and socket, a glass container such as a fish bowl, water, milk, a dark room, cardboard.

EXPERIMENT: Mount the lamp in the bottom of one can. Make a round hole in the bottom of the other can (a large nail can punch the hole) then tape the cans together as shown. Place the bowl of water where the light beam will shine through it, and add one or two drops of milk. A disc of cardboard with a hole in its center goes in the end of one of the cans.

The beam of light seen in the water is called a Tyndall cone, after John Tyndall, a British scientist, although the effect was first seen by Faraday in 1857.

REASON: The light is reflected or "scattered" by the tiny colloidal particles from the milk, and is polarized. Various sizes of particles give various colors to the cone, which is only slightly cone

The Tyndall Effect

shaped after all, but is extremely interesting to a technical man. Tyndall made the first exhaustive study of it. If the particles are smaller in diameter than one-twentieth the wavelength of light, blue is the predominant color.

The definition of Tyndall Effect is: visible scattering of light along the path of a beam of light as it passes through a system containing discontinuities. (McGraw-Hill *Encyclopedia of Science and Technology*.)

Without the cardboard disc between the cans the light does not come out in a narrow beam but reflects many ways from the insides of the cans. Spraying the insides of the cans with dull black will eliminate the need for the cardboard disc.

MACH BANDS

NEEDED: A shaded lamp, white card, another card, any color, a darkened room.

EXPERIMENT: Hold the card as shown so the light falls on half of the white card. Notice that where the shadow begins there is a light streak next to the bright reflection from the card, and a dark streak on the dark shadow side of the card.

REASON: Of course there are really no such lines; this is an example of how the eyes and nervous system can fool our brains.

But if the experiment is done correctly the imaginary lines are very clear.

The Austrian physicist, philosopher, and psychologist, Ernst Mach, first reported these bands to scientists a hundred years ago, and formulated a principle for the effect. A technical article touching on Mach bands appeared in *Scientific American,* June 1972.

Try seeing the bands around your shadow on concrete on a sunny day. Sometimes they are more clearly seen if the body is moved.

Mach Bands

POLARIZED LIGHT

NEEDED: A pane of window glass, Polaroid sunglasses, a light source such as sky or clouds.

EXPERIMENT: Hold the glass at about the angle shown in the drawing. Look at the glass through a sunglass lens, and rotate the lens, all the time experimenting with different angles of the window glass. A position will be found where light reflected from the sky or clouds is almost entirely cut off.

REASON: When the glass pane is held at the correct angle to the incoming light it takes out nearly all the light except that vibrating in one direction—that is, it polarizes the light. The Polaroid sunglass lens does the same thing as light is viewed through it.

The light reflected from the pane can be seen through the lens when the lens is in position to allow the polarized light from the pane to pass through it. If the lens is rotated 90 degrees it cuts off the polarized light from the pane.

A good encyclopedia gives lengthy explanations of this phenomenon.

Polarized Light

WATER STARS

NEEDED: A bright shaded light, medicine dropper, water, darkness.

EXPERIMENT: Shine the light down a stair well, and drop water from the light down (place a towel below to catch the drops). Reflections from the drops will shine like stars.

REASON: Light hitting a drop reflects in all directions from the surface of the drop, and is refracted in all directions as it passes through. The drop might be considered to be a million reflectors and a million lenses.

A few of the ways in which the drop reflects and refracts the light are shown in the diagram of a drop, which need not be perfectly round.

Water Stars

STROBE-LIKE LIGHT

NEEDED: A fluorescent light.

EXPERIMENT 1: With a fluorescent light as illumination, move the hand up and down rapidly. More than four fingers and one thumb appear to be seen.

Strobe-Like Light

REASON: This is a case in which the fluorescent light is a stroboscope. Its light goes dim then bright again sixty times per second, reversing itself 120 times each second.

When the light is bright we see the fingers. Then while the light is dim the hand moves to another position and we see the hand again when the light brightens, but in a new position. Retention of the view in both one position and the other gives the appearance of more than four fingers and one thumb. This retention is a trick of the eyes.

EXPERIMENT 2: Hold the hand in front of the television screen. The same effect is seen, and for the same reason. The TV picture goes on and practically off rapidly.

EXPERIMENT 3: Try this in front of a home movie projector while it is operating. Here the picture goes entirely off then on again rapidly, perhaps 24 times a second.

FLUORESCENT LIGHT

NEEDED: A fluorescent light and an electric fan.

EXPERIMENT 1: Watch the fan as it speeds up when turned on and as it slows down when turned off. It seems to go forward then backward and vice versa.

REASON: A fluorescent lamp does not burn steadily, but gets bright then dimmer 60 times a second as the current in the line

Fluorescent Light

reverses itself 120 times a second. The illusion is a stroboscopic effect and is more noticeable in some of the older lamps.

If the fan turns at such a speed that the blades are always in one position when the lamp is brightest then the fan seems to be stopped. If the fan turns a little more rapidly the blades seem to turn forward slowly. If the blades turn a little slower they will seem to be turning backward slowly because of the position of the blades when the light from the lamp is at its brightest point.

The human eye does not notice the fluctuation in the brightness of the fluorescent lamp.

EXPERIMENT 2: Run the fan in a room where the TV set is on. Turn out all other lights. The same effect is noticed and for the same reason. Also, the fan in the room where motion pictures are shown will produce the stroboscopic effect.

PINHOLE VISION

NEEDED: Cardboard, a pin, a stick four or five inches long.

EXPERIMENT 1: Hold the stick (the author used a Popsicle stick) against a printed page, the other end against the forehead. Close one eye and try to read the fine print. It may be impossible. Hold the card so you look through a pinhole in the card with one eye. The print may be clear.

REASON: The eye cannot normally focus at such a short distance. But when all the light reaching the eye comes through a tiny hole, focusing of the eye lens is not necessary. This is because the pinhole allows most of the light to pass through the middle portion of the eye directly to the center of the retina which is most sensitive. No refraction is required.

The diagrams should help to explain.

EXPERIMENT 2: Try pinhole vision if glasses are needed for clear distant seeing. Look through the pinhole—without the glasses.

LIGHT WAVES

NEEDED: Screen wire, a candle or bare electric bulb in a dark room.

EXPERIMENT: Look at the light through the screen. Streamers of light will be seen at right angles crossing at the light source. The lamp must be across the room rather than close to the screen for this effect.

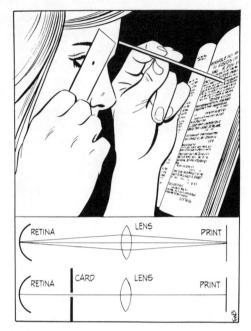

Pin Hole Vision

REASON: Each little hole in the screen acts as if it were a source of light, and the light waves come through in step (in phase). But when they come together in the eye their different path lengths vary and they are not all in phase. They act to cancel each other out

Light Waves

or reinforce each other adding to their brightness. This is called an interference phenomenon. Look up Fresnel's rule.

Try looking at the light through cloth, and also through two pieces of screen wire. Turn one of the pieces of screen as you look.

A RAINBOW PUDDLE

NEEDED: Dark colored bowl, water, oil.

EXPERIMENT: Moisten a toothpick or match stem with oil, and touch it to the middle of the water surface in the bowl. Look at the surface from various angles. Beautiful rainbow colors will be seen, as often seen in a puddle of water on a street.

REASON: Light rays striking the water surface are reflected in two ways: some from the surface of the oil and others from the surface of the water under the oil. The distance between the oil and water surfaces is such that "interference" in the two rays results. The interference serves to cut out some colors, allowing others to be seen.

As the oil film moves, its slight wedge shape allows different colors to show at different parts of the film. See diagram.

The author found that Three-In-One Oil gave best results in this experiment, although some other oils may be used.

A Rainbow Puddle

Mystery Colors

MYSTERY COLORS

NEEDED: An electric fan and a fluorescent light.

EXPERIMENT: Hold the fan so the fluorescent tube is reflected from one of the blades. Start the fan, and watch carefully. Colors may be seen in the reflection.

REASON: Phosphors inside the fluorescent tubes glow in red, green, and blue, but the colors are so mixed that the result is white light. Each of the phosphors has its own time of build-up of brightness and decay, and when the fan blades are turned at a specific speed they "tune in" to the time of one of the phosphors. At another speed they "tune in" to the time of another.

This effect is not easy to see. Only at certain definite speeds are the different colors seen; the colors change as the fan slows or speeds up. The strobe effect also is seen in the fan blades; they appear to stop and reverse themselves as the speed of the fan changes.

SUN'S RAYS

NEEDED: Observation.

OBSERVATION 1: The sun, on a foggy morning, shines through the trees in a very beautiful pattern as shown in the drawing.

Sometimes it makes a similar pattern shining through the clouds, and the old-timers explained this as "the sun drawing water."

REASON: The phenomenon occurs as the sun shines into and through tiny droplets of water (in the sky the water may be frozen). The light reflects and refracts in almost every direction. Some light from each droplet may reach the eye.

Sunlight shining on water droplets from a different angle makes the rainbow.

OBSERVATION 2: Note that the rays radiate apparently from a point just beyond one of the trees and extend from that point in all directions. We would expect the rays to seem parallel. Is this a matter of geometry?

Sun's Rays

THE TELESCOPE

NEEDED: Two magnifying glasses, one larger than the other.

EXPERIMENT: Hold the large glass at arm's length, and adjust the distance of the other from the eye so that a distant object, seen through both lenses, seems close.

REASON: This shows the principle of the refracting telescope. The objective lens forms an inverted real image of the object

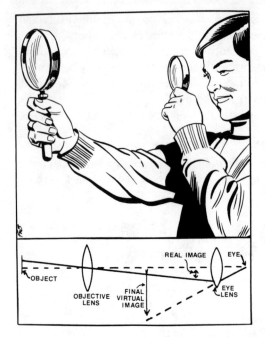

The Telescope

looked at. The eye lens views the real image of the objective lens and shows it to the eye as an enlarged and inverted virtual image.

A real image can be shown on a screen. A virtual image cannot. Reflection in a mirror is a virtual image; you cannot touch it or put it on a screen. Your face on the movie screen when your vacation movies are shown is a real image.

Most large telescopes are the reflecting type mainly because lenses cannot be made as large as mirrors. Purpose of the large size of the mirror or objective lens is not to magnify, but to gather light. The larger it is the more light it gathers.

TRICKY COLORS

NEEDED: Red, blue, or green cellophane, crayons the same colors, white paper. Cellophane must be transparent.

EXPERIMENT: Draw something on the white paper using the red and either blue or green crayons in the same drawing. Look at the drawing through the red cellophane, and the red lines almost disappear; look at it through the other cellophane, and only the red lines show distinctly.

REASON: Light from the sun or a lamp contains the primary colors, and so the red lines can reflect red from the original light, and the blue or green can reflect their colors. White light from the paper is reflected, but as it passes through the red cellophane nearly all of it is absorbed except the red. This makes the white paper and red lines appear near the same color of red.

It is the same for the other colors. The blue and white reflect but all the colors in the white are absorbed by the blue cellophane except the blue. Practically no blue is reflected from the red lines, and so they appear black. When the drawing is seen through the red cellophane the blue or green lines appear black because they do not reflect red.

REFLECTION UNDER WATER

NEEDED: A jar of water and a lighted candle.

EXPERIMENT: Place the candle behind the water jar as shown in the diagram. Look upward at the water surface, and the candle will be seen reflected, upside down.

REASON: We are accustomed to seeing trees and cliffs reflected from the upper surfaces of calm lakes. The under surface of still water reflects in the same way. Skin divers can see this phenomenon if the surface of the water in which they are swimming is sufficiently calm.

The magazine *Popular Science* suggested this phenomenon as a possible explanation for some of the sightings called unidentified flying objects. Automobile or other lights can reflect from the bottom of an atmospheric "layer" to viewers at a distant place.

Reflection Under Water

RAINBOWS IN A BUBBLE

NEEDED: Bubble blowing equipment.

EXPERIMENT: Light is reflected from both front and back sides of the film that makes the bubble. Light reflected from the inside surface reaches the eye a little later than that reflected from the outside surface, because it has a slightly longer path.

Because of what is called "interference" some of the rays reach the eye in phase, and are bright, while others are out of phase and tend to cancel each other out. Varying thickness of the bubble film accounts for different colors. When the thickness is very small the waves all cancel each other, and the appearance is dark.

SODIUM LIGHT

NEEDED: Alcohol, salt, a jar lid, a match, a large pan or board, different colored objects.

EXPERIMENT: Put some salt into the lid, then pour alcohol into it. Light the alcohol, turn out the lights so the room is dark, then sprinkle salt down into the flame. Look at the different objects.

REASON: Rubbing alcohol is not good for this. The author tried methyl alcohol (wood alcohol), and a red box looked black. Fuel alcohol works, and its flame made the red box show as yellow. The difference is in the chemicals in the alcohol itself which add their colors to the flame. A pure sodium flame, from the sodium in the salt, would be pure yellow with no other colors.

DRIVING IN THE FOG

EXPERIMENT: When headlights are turned to the bright or "up" position it is more difficult to see the road through the fog.

REASON: When a light ray hits a fog droplet, the light is reflected and refracted in all directions, so that the droplet would have the appearance as shown in the bottom right drawing.

The drawing of an enlarged fog droplet, at bottom left, shows some of the ways in which the light rays can reflect and refract when they hit the droplet. The same ray is reflected and refracted many times, as it hits many droplets, giving an overall dispersion or scattering of the light. When the automobile beams are turned down the light hits fewer droplets, so that the fog is less bright and the road is seen more clearly.

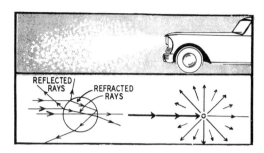

RAINBOWS IN A RECORD

NEEDED: One long-playing phonograph record.

EXPERIMENT: Hold the record so that light coming toward you shines on the record grooves. Rainbow colors may be seen.

REASON: The light is reflected from the irregular surfaces of the needle grooves, causing interference of the reflected light. This phenomenon is similar to that seen when light is reflected in rainbow colors from a soap bubble. The bubble rainbow was explained in a previous experiment.

A Candle Flame Shadow

A CANDLE FLAME SHADOW

NEEDED: A candle, a white or gray card, sunlight.

EXPERIMENT: Hold the candle so that its shadow is cast on the card. The shadow of the flame may be seen, but rather faintly.

REASON: The flame is composed largely of carbon particles in varying stages of burning. The particles do not look black in the yellow flame, but they are there and can make a faint shadow.

Also, what we see as a shadow is partly the refraction of the light as it passes from cool air through the hot flame gases and back again into cool air. Light is refracted when it passes from one medium to another of different density, and the hot gases of the flame are less dense than the cool surrounding air.

Chapter 12
Household Hints

QUENCH THE FIRE

EXPERIMENT: Sprinkle baking soda on a grease fire to extinguish the flames

REASON: Baking soda is $NaHCO_3$. When two molecules or any multiple of two are heated, they break up into three fire-killers, as follows: $Na_2CO_3, + H_2O + CO_2$.

In other words, the soda becomes sodium carbonate, a solid which coats the burning grease and helps prevent burning, water which cools the burning grease, and carbon dioxide which helps smother the flames by cutting off the supply of oxygen. This household hint works!

CLEAN COPPER OR BRASS

NEEDED: Household ammonia.

EXPERIMENT: To clean copper or brass, soak in household ammonia, full strength, until the metal is clean. Polish with a cleaning pad of the commercial variety made from very fine steel wool and soap. Do not use regular steel wool; it scratches.

REASON: The corrosion or blackening of the copper or brass is made up of both oxide and dirt or smoke. These do not dissolve readily in water.

The ammonia forms a complex "ion" with the corrosion—a charged group of atoms that will dissolve in water and so may be washed away. The copper ammonia ion is bright blue. Observe the development of this color while the copper is in the "household ammonia" solution.

THOSE DARK STREAKS

PROBLEM: In the usual picture-framing job, sealing tape or paper is pasted over the back of the frame and the picture backing. Smoke and fine dust can enter between the picture and the glass as shown by the arrows in the upper drawing. The sealing actually does not seal anything.

AN IMPROVEMENT: Place the picture, the glass, and the backing on the edge of a table, and seal them together with cellophane tape as shown in drawings 1, 2, and 3. If care is taken, even the corners can be rather well sealed in this way, so that no dust and smoke can enter except through the pores of the picture backing.

This eliminates much of the dark streaking on pictures that hang for long periods of time on the wall.

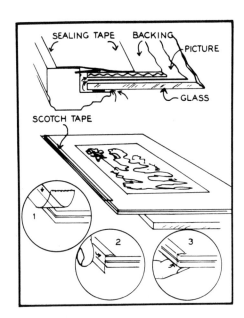

Those Dark Streaks

PICKLES

NEEDED: Fresh cucumbers, salt water, alum water.

EXPERIMENT: Soak the sliced cucumbers in salt water, and they will get very limber, because the salt water draws water out of the cells of the pickles by osmosis. This is the process by which a less concentrated solution will flow through a membrane into a more concentrated solution.

Then soak the slices in the alum water. The slices will again become firm.

REASON: The alum solution is a very dilute solution. Water passes into the cells of the cucumber producing a firmness known as "osmotic turgescence."

BUGS AWAY!

EXPERIMENT: When washing greens the tiny bugs wash off much better if salt is added to the water.

REASON: Salt is objectionable and irritating to most insects and other small animals. Most bacteria cannot live in a salt solution, therefore salt can be used to preserve food, by preventing spoilage from bacteria.

Salt solutions coming in contact with cell walls can draw water from the cells by exosmosis, and thus kill the cells. This can be felt in the mouth after eating something very salty. The lips seem to "draw" as their cells lose water.

A WATER RACE

NEEDED: Two pieces of cloth and two small puddles of water.

EXPERIMENT: Wet one of the cloths and squeeze it out. Place both cloths down on the puddles, and it will be seen that the damp cloth soaks up water much faster than the dry one.

REASON: The dry cloth has a considerable air film which prevents the entrance of water. The partly damp cloth has no such air film and entrance to the pores of the cloth is much more rapid as a result.

POT HOLDERS

NEEDED: Two pot holders.

EXPERIMENT: Pick up a hot pan with a dry pot holder and no burn is felt. Pick up the same pan with the wet pot holder and the hand is likely to be burned.

REASON: A dry pot holder consists of fibers with much insulating air space between. Air does not conduct heat very well; water conducts it much better. When the pot holder is wet, water takes the place of the air spaces and will conduct the heat through to the hand. It is possible, if the pan is very hot, that enough steam will be produced to go through the wet cloth to burn the hand.

DON'T CUT THE GLASS!

THE SUPERSTITION: An old-wives method for sharpening scissors consists of rubbing them on a glass as if trying to cut it.

CONCLUSION: The scissors may be sharpened somewhat if care is taken that the cutting edges rub against the glass in the correct way shown. If the cutting edges rub the glass incorrectly, the procedure will dull the scissors.

A more reliable way is to place the scissors in a vise and file them as shown in the lower drawing.

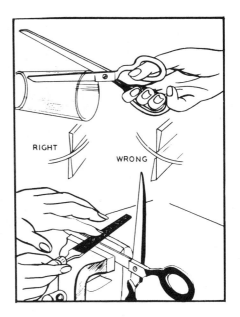

Don't Cut the Glass!

HOW TO CLEAN EYEGLASSES

NEEDED: Soap, running water, a freshly laundered cloth.

EXPERIMENT: Wet the glasses, then rub soap on the lenses with the fingers. Rinse the soap off carefully, hold to a light, and wipe them with the cloth.

OBSERVATION: If the lenses are rubbed before washing, oil on them will spread and leave the lenses coated or streaked with it. Also, gritty dirt can scratch the glass if it is not washed off before they are rubbed. This can be shown by putting some sand on a scrap of window glass and rubbing it.

Never lay the glasses down so that the glass touches a table top or other hard surface. Gritty dirt is certain to scratch them if this habit is allowed.

SOGGY PANCAKES

NEEDED: A hot pancake and a cold plate.

EXPERIMENT: Place the hot cake on the cold plate. Lift it after a few seconds; there will be water in the plate, enough to make the hot-cake soggy.

REASON: Heat drives water out of the dough in the form of vapor. After the cake is taken from the griddle the heat inside it is still driving out moisture, which condenses on the cold plate.

The best pancake bakers place the cakes on a hot plate. Moisture does not condense on the hot plate, but continues to evaporate into the air. The situation is one of drying the cake rather than wetting it.

Soggy Pancakes

LUMPY CUSTARD

NEEDED: The custard ingredients.

EXPERIMENT: When using egg yolks to thicken custard, beat the yolks, then add several tablespoonfuls of the hot custard to the yolks, and mix. This warms the yolks and is called "tempering." It must be done before the yolks are added to the custard to prevent the formation of lumps in the finished custard.

REASON: If egg yolk is heated too quickly, as by placing it in hot custard, it cooks into hard lumps as it would if placed in a hot pan. It is the protein that hardens in this way.

Lumpy Custard

PICKLED EGGS

NEEDED: Hard boiled eggs, pickling solution (containing salt and vinegar), a container.

EXPERIMENT: Peel the hard boiled eggs, and put into the solution. They float. As they soak up the pickling solution they start sinking to the bottom. When on the bottom they are pickled and ready to eat.

REASON: At first the density of the eggs is slightly less than that of the pickling solution, and the eggs float. But the solutions begin to diffuse—the pickling solution goes slowly into the eggs and the egg liquid goes into the solution. The result is that the density of the eggs increases so it is greater than that of the solution, and the eggs sink.

YOGURT ANYONE?

NEEDED: One quart of milk, three tablespoons of commercial yogurt, a warm oven.

EXPERIMENT: In the top of a double boiler heat the milk to a boiling temperature and then cool it to about 115 degrees F. Add three tablespoons of commercial yogurt, mix well, and pour into jars or custard cups. Place in a pan of water about 115 degrees and keep

Pickled Eggs

in a warm oven to keep water at the same temperature until milk sets. You may use the last of this batch to start your next one.

REASON: A special kind of Lactobacillus acidophilus is alive in commercial yogurt. A temperature of 115 degrees Fahrenheit is just

Yogurt Anyone?

right for them to grow and multiply, and it is these little bacteria that make the yogurt. They follow the exponential growth law: one turns to two in a certain time, the two turn to four in a similar time, four turn to eight, etc.

Heating the milk to boiling point kills off any bacteria that might multiply the same way and spoil the special work of the Lactobacillus acidophilus.

MAKE TRACING PAPER

NEEDED: Paper, gum turpentine, a brush.

EXPERIMENT: Place a sheet of paper on a piece of window glass or other flat surface, and brush turpentine over it. It becomes translucent, and can be placed over a drawing to be traced. When it dries it resumes its original state, but the traced drawing remains.

REASON: The paper consists of irregular strands of matted cellulose which scatter and reflect light so that dark lines of color or black cannot be seen through an ordinary sheet. The turpentine allows a temporary passage of light. Dark lines show dark and white areas are white. But the turpentine evaporates quickly.

Turpentine may be applied with a paper towel, but prolonged contact with the skin should be avoided.

Make Tracing Paper

MAKE FINGER PAINT

NEEDED: One package plain gelatin, a half cup of cornstarch, a half cup of mild detergent, food coloring, water.

EXPERIMENT: Mix the gelatin in a fourth of a cup of water, and set it aside. Mix the cornstarch with a three-fourths cup of

Make Finger Paint

water, add two cups hot water, and bring to a boil while stirring constantly. Let the mixture boil until it thickens, take away from the heat, and add the gelatin and detergent. Mix well, and pour into different jars where color may be added.

This is a harmless paint for little fingers. Keep it in closed jars for storage. Molds may grow in the gelatin after a time and the paint will be useless.

A WEED GOOD TO EAT

NEEDED: A lawn not too well kept.

EXPERIMENT: Look for the broad-leafed plantain weed, "plantago." It grows flat against the ground, its leaves growing directly from the root. Try to isolate one of the plants so it can grow a slender spike which will blossom into greenish flowers and later produce seeds. This life cycle is seldom noticed.

Many weeds and shrubs are poisonous. But plantain is good to eat, especially when the leaves are young and tender. The older leaves may taste bitter. The leaves may be boiled and served with butter and a little vinegar, or may be used in salads. They are good rabbit food, also.

A Weed Good to Eat

MAKE SOUR MILK

NEEDED: Vinegar, fresh or evaporated milk.

EXPERIMENT: When sour milk or buttermilk is needed in a recipe, and there is none, put one tablespoon of vinegar in one cup of fresh milk, and stir. It may be used as sour milk. Diluted evaporated milk may be soured in the same way.

REASON: Vinegar is about four percent acetic acid. Sour milk contains lactic acid, formed from lactose sugar (present in sweet milk) by action of bacterial enzymes. Either acid serves the same purpose in the cooking recipe, and both are desirable in food in the small amounts involved. Lemon juice, which contains citric acid, may be used.

Make Sour Milk

CHEMISTRY AND CURDS

NEEDED: Tomatoes to be cooked with milk or cream, soda.

EXPERIMENT: Cook tomatoes with milk or cream, and curdling will likely result. Try cooking them with milk or cream to which a pinch of soda has been added, and—no curdling.

REASON: The curdling is due to the acid in the tomato juice causing the protein of the milk to coagulate. Baking soda neutralizes the acid in the tomato juice and the protein does not coagulate in the absence of acid.

Fresh milk does not coagulate, but sour (acid) milk which results when lactose of the milk changes into lactic acid does coagulate.

BROWN CUPCAKES

NEEDED: Observation as cupcakes are baked in vari-colored paper cups.

OBSERVATION: The cakes baked in darker colored paper are browner.

EXPERIMENT 1: Dark clothes are warmer in winter—not quite an exact statement. Put on a light colored glove and a dark glove. In sunlight the dark glove will feel warmer; it absorbs more

Chemistry and Curds

energy. But in darkness it will be colder; it emits more energy. It does not have to be winter.

EXPERIMENT 2: Dark colored paper or cloth will melt through snow faster than light colored paper or cloth—as long as light energy is being absorbed by them. In darkness the dark colors will radiate or emit more energy and be colder.

EXPERIMENT 3: Note that ice will melt more quickly on dark pavement than on light pavement—in daytime when light shines. In the dark the darker pavement (with starting temperature the same as the light colored pavement) emits more radiation and can be colder.

REASON: Light colors reflect more light and heat than dark colors, and dark colors absorb more light, turning it into heat. Therefore, the cakes baked in dark paper get just a little hotter, and browner.

Brown Cupcakes

Double Boiling

DOUBLE BOILING

NEEDED: A double boiler, a candy thermometer, water, a stove.

EXPERIMENT: Get the double boiler steaming, and take the temperatures in upper and lower vessels. The author, using this crude method, obtained readings of 208 degrees in the lower vessel and 200 degrees in the upper vessel.

Add three heaping teaspoonfuls of salt to the lower vessel, let it steam, and take the temperatures again. The author's readings this time were 216 degrees in the lower vessel and 204 degrees in the upper.

REASON: Adding salt raised the boiling point of the water in the lower vessel, and at the same time the temperature was raised in the upper vessel. (Incidentally, salt added to water also lowers its freezing point.)

In the author's test four cups of water were used in the lower container and two in the upper container.

GELATIN AND EGG

NEEDED: Gelatin, an egg

EXPERIMENT: Note that gelatin is a solid at room temperature and a liquid when heated. It turns back to solid when cooled. Remember that egg white is liquid at room temperature and becomes solid when heated. The egg remains solid when cool.

REASON: Egg coagulates when it is heated because small particles join one another to form a mass that remains together whether hot or cold. Gelatin is generally dissolved in a limited amount of water and heated mildly. When cooled this liquid may become a colloidal solid with the water as a part of the distributed total.

The water is not added to the egg but what is naturally present forms a permanent suspension when heated to the boiling point of water.

The clotting and setting of egg albumen is irreversible. The solidification of a gelatin solution is reversible.

Look up "colloid" and "suspension."

Gelatin and Egg

Remove Hardened Wax

REMOVE HARDENED WAX

NEEDED: A deep freeze, or dry ice; blotting paper, a hot iron.

EXPERIMENT 1: When wax drops on a fabric and hardens, place the garment in the deep freeze for an hour or two. While it is still very cold, bend and break and rub the spot and the wax may turn to powder and brush off.

EXPERIMENT 2: If dry ice is available it is better because it is colder, but care must be taken in the handling of dry ice (solid carbon dioxide).

EXPERIMENT 3: If these methods do not work or are not practical, place the blotting paper over the spot of wax, and heat with the iron. Press firmly, and perhaps move the blotting paper around. The wax should be softened enough to flow up into the blotting paper, leaving the cloth.

COMMENT: The flow of the melted wax up into the fabric is by capillarity.

Dry ice can harm the skin. Hold it with several layers of cloth or with tongs.

Dry ice will remove stuck-on chewing gum, even if in someone's hair!

HOW FRESH THE EGG?

NEEDED: Eggs.

EXPERIMENT 1: Put the eggs into a pan of water. If an egg is fresh it will rest on its side on the bottom of the pan. If it is a few days old one end will turn upward. If it is stale it will stand on one end, and if rotten it will float.

REASON: As the egg gets older some of the white and yolk lose water through the pores of the shell, and the liquid is replaced by air. The air accumulates in the large end of the egg, where there already was an air cell.

The air is buoyant and tends to lift the egg, more and more as the egg gets older.

EXPERIMENT 2: To tell a hard boiled egg from a raw egg: spin the egg on its side. The cooked egg will spin merrily, because the liquid inside the shell has been hardened and turns with the shell. When the raw egg is spun the liquid does not all spin with the shell, and actually acts as a brake to stop the spinning of the shell.

How Fresh the Egg?

VITAMIN CHANGE

NEEDED: Orange juice, corn starch, tincture of iodine, a cooking pan and heat, water, glasses, spoons, eye dropper.

Vitamin Change

EXPERIMENT: Bring orange juice to a boil. Put uncooked juice into a container. Boil a half teaspoon of starch in half a glass of water, and put 20 drops of it into each of two glasses of water. Put two drops of iodine in each mixture, and stir. The color should be blue. Add 25 drops of heated juice to one glass, and stir. Then see how many drops of the other juice are necessary to produce the same color, in the other glass.

REASON: A mysterious combination of starch and iodine produces the blue color, and vitamin C destroys the color. This experiment should show that heating the juice destroys some of the Vitamin C. Less unboiled juice should be needed to remove the color.

It is not customary to boil orange juice, but for the purposes of this experiment the juice is easier to handle than apples or peas or other foods rich in vitamin C.

Boiling a solution of vitamin C destroys it. It is not the heat that does the job, however; it is the oxygen of the air that destroys the vitamin C, and boiling simply speeds up the action.

Chapter 13

Chemistry

A TRICK IN VOLUME

NEEDED: Salt, water, a glass, a dinner plate.

EXPERIMENT: Fill the glass full of water, place it on the dry plate, then pour in a heaping tablespoon of salt. Stir until the salt is dissolved.

Some water will spill out on the plate. After the salt is dissolved, pour the water from the plate back into the glass. It may be possible to pour it all back without spilling any. Or if some spills it will be very little, not nearly as much as the volume of salt dissolved.

REASON: When the salt is dissolved in the water, the molecules of each crystal fit into spaces around each other much as sand would fit in the spaces between marbles in a jar. Much air is between the dry salt crystals but it does not enter the solution. The particles of the salt in solution are about one twenty-five millionths of an inch in diameter.

"BAD BREATH"

NEEDED: Two identical jars, two candles, a soda straw, a match, a pan of water.

EXPERIMENT: Fill a jar with water, insert it in the pan of water, with the neck down, then raise it up so that air from the room may enter and the water may run out. Place the jar on the table, neck down.

Fill the next jar with water the same way, but blow breath under it with the straw to force out the water. In doing this, take a deep breath, so that the air through the straw will come from the lungs, and not just from the mouth. Place the jar neckdown beside the first.

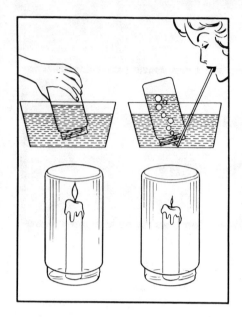

"Bad Breath"

Light both candles, and place the jars over them at the same time. The candle will burn longer in the jar containing room air.

REASON: The candle flame goes out when the oxygen in the jar is used up. Room air contains about 20 percent oxygen. Exhaled air contains about 16 percent oxygen and 4 percent carbon dioxide. The oxygen not only has been partially replaced with carbon dioxide, but the carbon dioxide interferes with the burning of the candle.

GROW CRYSTALS

NEEDED: Water, Epsom salts, a piece of glass.

EXPERIMENT: Dissolve some of the salts in the water. Pour a few drops of the liquid on the glass, and let it evaporate in a warm place. The salts will be left in the form of beautiful crystals. Examine them under a magnifying glass or a microscope.

OBSERVATION: Exactly why the molecules arrange themselves in definite patterns to form crystals is not known. The shape of the crystal depends upon what substances are used and what the conditions are.

Try other salts, such as table salt or photographer's hypo. Snowflakes are examples of crystals. Note that no two crystals are alike. Some unusual crystal forms may be seen if an egg is broken and its white and yellow allowed to evaporate on a dish.

Crystal formation requires time and room.

A SALT GARDEN

NEEDED: A dish, water, salt, vinegar, some small porous stones or pieces of coal.

EXPERIMENT: Place the stones in the dish. Pour salt into warm water and stir until no more salt will dissolve. Put a spoonful of vinegar in the water, and pour it over the stones in the dish. In a few days, the salt will begin to "grow" and eventually will cover the stones in beautiful crystals.

REASON: The salt water flows up through and over the stones because of capillary action, and as it rises it evaporates leaving the salt behind. The vinegar serves to take away oily spots on the stones that would interfere with the free upward flow of the salt water.

If the solution is left in an open glass the crystals will begin to form above the solution. Crystals there will act as capillaries, allowing more liquid to flow up, more crystals will form, allowing more liquid to flow up. This may continue until there is "growth" of the crystals up and over the edge of the glass.

SEPARATE SALT AND SAND

NEEDED: Salt, sand, water, a cooking vessel, a stove or hot plate.

EXPERIMENT: Mix the salt and sand. Dissolve the salt in hot water. The sand will settle to the bottom of the container, so that the

Separate Salt and Sand

water containing the salt can be poured off. Boil away the water, and the salt will remain.

REASON: The sand does not dissolve in the water as the salt does. The salt does not boil away as the water does.

Will this work with a mixture of sugar and sand? What about sugar and salt?

THE SWOLLEN EGG

NEEDED: An egg, vinegar, water.

EXPERIMENT 1: Soak the egg in vinegar until the shell has become soft. Pour out the vinegar, and soak the egg in plain water. In a few days the egg will have become so large that the shell will burst.

EXPERIMENT 2: Without the vinegar treatment, peel just a little square of eggshell off the egg, without breaking the membrane below. By osmosis, the water will go into the egg, pressing upward on the exposed membrane until it breaks.

REASON: The acid in the vinegar combines with the calcium in the egg shell, making the shell soft. When the egg is then immersed in water, the water will go through the shell into the egg until the shell bursts. This is the principle of osmosis, in which a less concentrated solution will go through a membrane into a more concentrated solution more than in the reverse direction.

SALT AND VINEGAR ON COPPER

NEEDED: A copper cent, vinegar, salt, a plate.

EXPERIMENT: Sprinkle a little salt on the coin, pour on some vinegar, and the coin will be cleaned beautifully.

Let it stand for a while, and the copper coin begins to corrode and turn green.

REASON: Here is a little simple chemistry. Salt, or sodium chloride, plus hydrogen acetate in the vinegar, gives us sodium acetate plus hydrogen chloride, or hydrochloric acid.

The hydrochloric acid is strong, and if the bronze cent containing 95 percent copper, 3 percent zinc, and 2 percent tin comes into contact with it and the newly formed salt (sodium acetate), it is cleaned rapidly.

The cleaning process leaves the surface in porous "active" condition so that it quickly corrodes by combining with water, and oxygen and carbon dioxide in the air.

TECHNICOLOR CABBAGE

NEEDED: A cooked red cabbage and various substances for testing.

EXPERIMENT 1: Pour reddish-purple juice from the cooked cabbage into various containers. It will change its color to reveal whether a substance is acid or base.

Lemon juice, which is acid, will turn the solution pink. Baking soda, which is base, will turn it green. Try other substances.

EXPERIMENT 2: Try beet juice as an indicator. Bases will turn it blue.

EXPERIMENT 3: Laundry bleach will take the color out.

REASON: All acids contain hydrogen ions which change the color of litmus and other indicators, including red cabbage juice. Bases contain an oppositely charged negative hydroxyl ion which produces an opposite color effect on litmus and some other vegetable colors including red cabbage juice.

A JAM EXPERIMENT

NEEDED: Blackberry jam or jelly, warm water, substances to test.

A Jam Experiment

EXPERIMENT 1: Do this: Put a spoonful of the jam or jelly into a glass of warm water, dissolve it, and the color is probably red. Put a few drips of ammonia into it, and the color changes to greenish purple.

EXPERIMENT 2: Add just enough ammonia to produce the greenish purple color. Then add lemon juice or vinegar, which are weak acids, and see the color change back to red.

REASON: Many colors from flowers and vegetables can serve as indicators, to tell whether a substance added is acid or alkali. The jelly solution is red when an acid is added to it, greenish purple with an alkali.

The red color of the jam is a natural indicator. An excess of H ions furnished by an acid gives one color. An excess of (OH) ions furnished by any base gives the other color.

EASY OXYGEN

NEEDED: Hydrogen peroxide (20 percent solution from the pharmacy) and an old dry cell.

EXPERIMENT 1: Put the hydrogen peroxide into a jar (two tablespoonfuls), and sprinkle a tablespoonful of the black mixture from the dry cell over it. Notice the large quantity of bubbles—bubbles of oxygen.

OBSERVATION: The manganese dioxide in the black mixture combines with the hydrogen peroxide to produce manganese hydroxide and free gaseous oxygen, according to the equation $MnO_2 + H_2O_2 \rightarrow Mn(OH)_2 + O_2$.

MnO_2 is also a catalyst to speed up the decomposition of the unstable H_2O_2 which decomposes as follows:

$$2H_2O_2 \xrightarrow{MnO_2} 2H_2O + O_2$$

EXPERIMENT 2: Test to see if oxygen is in the jar by placing a glowing splinter in the jar. The oxygen will make the splinter burst into flame.

SPACE PROBLEM

NEEDED: A small glass, a larger one, a wax marking pencil, some water, some rubbing alcohol.

EXPERIMENT: Fill the smaller glass full of water twice and pour the water into the larger glass. Mark carefully on the glass where the surface is. Empty the big glass.

Now fill the smaller glass once with the rubbing alcohol and once with water and pour into the large glass. Be careful each time to fill the smaller glass to the same degree of fullness. The surface will be below the mark, showing that the mixture does not occupy as much space as the water.

REASON: The molecules of water and alcohol are not the same size, and so the smaller molecules can fit somewhat into the spaces between the larger ones. An experiment to demonstrate this: pour equal volumes of marbles and sand into a container, and see how the sand fits into the spaces between the marbles.

BURN FLOUR

NEEDED: Cheesecloth, flour, a lighted candle.

EXPERIMENT: Make a bag by folding a double thickness of cheesecloth, put flour into it, and shake it over the candle flame. The flour will burn in sudden, sparkling flames.

REASON: Flour does not burn readily unless the particles are so separated that oxygen in the air can reach every one. This condition is met as the fine flour dust comes down to the flame. The finer the flour, the more surface is available for this rapid oxidation.

When the proportion of flour particles to air is just right in a large area, an explosion powerful enough to wreck a flour mill can result. The small explosions in this experiment are harmless, except that fire is always dangerous to some extent. Be careful.

Burn Flour

A "POP" BOTTLE

NEEDED: A bottle with a cork, water, vinegar, baking soda, paper, a pan or sink to catch the overflow.

EXPERIMENT: Put water and vinegar into the bottle. Roll a little baking soda in the paper (tissue is best). Drop the paper into the bottle, put the cork on, and soon the cork will pop out.

REASON: When the acid of the vinegar and the baking soda are allowed to mix, they combine chemically to produce carbon dioxide gas. The pressure of the gas as it is formed blows out the cork.

Corks may be hard to get. Try this without the cork; the water and bubbles will overflow. Do not use a screw-top bottle.

A SIMPLE INDICATOR

PROBLEM: Make an Indicator.

NEEDED: Slices of fresh beet, water, a pan, heat, a drinking glass.

EXPERIMENT: Put half a glass of water into the pan. Put in three or four slices of beet and boil for five minutes. Let the red liquid cool, and pour some of it into the glass. Test various substances with it to see whether they are acid or alkaline.

If alkaline substances such as alum water or ammonia are added the color turns to yellowish green or brown. If an acid such as lemon juice is added the reddish color returns.

REASON: Indicators are natural colors from plants which show one color with an excess of H^+ ions from an acid and a different color with an excess of $(OH)^-$ ions from a base. Litmus is the most common indicator in laboratories. The exact mechanism of the color change is not thoroughly understood.

A Simple Indicator

MAKE CARBON DIOXIDE

NEEDED: A gallon jug, a beach ball, a cork stopper, a medicine dropper, water, sugar, molasses, yeast.

EXPERIMENT: Mix a cup of sugar, two tablespoons of molasses, and yeast in 3/4 jug of water. Set in a warm place, attach the ball to it after squeezing out the air. Overnight the ball should fill with carbon dioxide.

REASON: Yeast plants growing in the solution change sugar to alcohol and carbon dioxide gas. The gas fills the ball, as it filled the balloon, with this difference: pressure created in the balloon stopped the fermentation process, and the balloon did not expand. The ball could fill because there was no pressure from tight rubber.

Hold the opening of the ball to the mouth or nose and "taste" some of the gas. It may smell like molasses, and will definitely have a tingle as from a soft drink. Pour some of the gas into a jar, lower a candle into it, and the candle flame goes out.

Inset drawing shows how the cork and medicine dripper are placed to fill the ball.

Make Carbon Dioxide

CARBON DIOXIDE BY FERMENTATION

NEEDED: Gallon jug, yeast, sugar, a balloon.
EXPERIMENT, Dissolve yeast and sugar in a gallon of water in the jug. Fit the rubber balloon over the mouth of the jug, and set

Carbon Dioxide by Fermentation

the experiment in a warm place where the light is not bright. Bubbles will form in the liquid. They are carbon dioxide, and will have enough pressure to partially inflate the balloon. This takes several days.

REASON: Sugar, either $C_6H_{12}O_6$ or $C_{12}H_{22}O_{11}$ in the presence of zymase, a complex of enzymes, from the yeast plants, gives alcohol and carbon dioxide CO_2. The alcohol dissolves in the water while the gas escapes the water and goes into the balloon.

LIGHT AND CHEMISTRY

NEEDED: Construction paper or newspaper, sunlight.

EXPERIMENT: Place the paper in the sun, and place small objects on it. Leave it for several hours or days. The paper will have changed color where the light hit it. Outlines of the objects may be seen.

REASON: Paper has a natural dye which is very pale or it may be treated with coal tar dyes to give brighter color. The dyes disintegrate and the color of the paper changes slowly when oxygen of the air is activated by bright sunlight. Not all chemical changes resulting from light are well understood.

Light and Chemistry

CHEMICAL ACTION

NEEDED: A candle flame or other fire.

EXPERIMENT: Note that heat is produced as a substance is burned.

REASON: A chemical substance such as candle wax has a certain amount of energy per unit of weight. When a chemical action takes place the products of the action may have more energy or less energy than the original substances.

A "fuel" is a substance which burns by combining with oxygen of the air to produce products which possess less energy per unit of weight than the original fuel. The difference is liberated as heat, light, and/or motion energy.

When a fuel is burned and the energy is liberated the products are generally worthless and may be dangerous. Automobiles use the energy liberated when gasoline is burned, and the worthless products escape through the tailpipe and pollute the air.

BLEACH FLOWERS

NEEDED: Flowers of different kinds, a jar, a string, a plastic lid for the jar, household ammonia.

Chemical Action

EXPERIMENT 1: Tie the flowers and suspend them in the jar by running the string through the lid. Pour a little ammonia into the jar and watch colors disappear.

Our red, pink, and purple flowers turned green. White and yellow flowers were not changed.

REASON: Substances that give the petals their color are present along with chlorophyll which is green. The ammonia gas which rises from the household liquid destroys some of the colors but not others. When colors that hide the green chlorophyll are destroyed the chlorophyll color shows.

EXPERIMENT 2: Try pieces of colored paper and cloth, wetting them before placing them in the ammonia vapor. See how many of their colors are changed.

Bleach Flowers

POLLUTION TO ORDER

NEEDED: Photographer's hypo, lemon juice, an old spoon, a candle for heat.

EXPERIMENT: Put a few crystals of hypo into the spoon with a little lemon juice. Heat it and smell cautiously. The unpleasant choking odor is sulphur dioxide, which is one of the undesirable air pollutants coming from some industries.

REASON: Hypo is sodium thiosulphate, essentially $Na_2S_2O_3$ and when heated with lemon juice some of the sulphur (S) and oxygen (O) atoms unite as the hypo molecules are destroyed by the heat. The sulphur dioxide (SO_2) formed always has a stinking, unpleasant odor, and is poisonous, but the amount produced in this experiment is not dangerous.

A BIG BAD ODOR

NEEDED: Photographer's hypo (sodium thiosulphate), an old spoon, a candle for heat.

EXPERIMENT: Put a few hypo crystals in the spoon and hold it over heat until the hypo melts. Continue to heat it, and the odor of rotten eggs will be given off. The odor is that of hydrogen sulphide, and this is a simple chemical experiment.

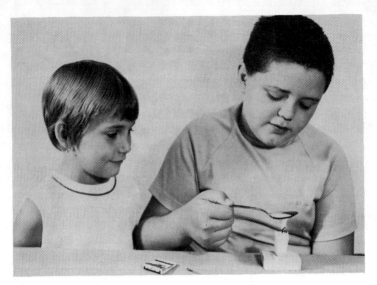

Pollution to Order

REASON: There is some water of crystallization in the seemingly dry hypo crystals. When the crystals are heated hydrogen (H) atoms from the water and sulphur (S) atoms from the hypo unite to form hydrogen sulphide (H_2S) molecules.

Rotten eggs also produce H_2S molecules, which always smell the same regardless of how they are produced.

CASE OF THE DISAPPEARING CRYSTALS

NEEDED: Salt or sugar and water.

EXPERIMENT: Dissolve a spoonful of the crystals in a glass of water. They disappear; the water becomes clear. When the water evaporates, the crystals again appear. Where do they go when they dissolve?

REASON: Crystals consist of thousands to millions or more of molecules of a definite chemical composition held together in a definite geometric pattern. When they dissolve they separate into smaller or individual invisible particles distributed in the solution.

When the solution is evaporated the smaller particles again unite in a definite geometric pattern if enough time is allowed, and they become large enough to be seen again.

A FIRE HAZARD

NEEDED: A lighted cigarette and dry grass or pulverized dry leaves.

Case of the
Disappearing Crystals

EXPERIMENT 1: Try to set fire to the dry material by dropping the cigarette into it. It is difficult.

EXPERIMENT 2: Try lighting shavings from a pencil sharpener with a cigarette.

EXPERIMENT 3: Place a lighted cigarette and a lighted cigar on an ash tray. The cigarette will continue to burn; the cigar will not. Cigarettes actually are incendiary bombs. They can set fire to grass and leaves in the open when there is a breeze.

A Fire Hazard

Oxygen

OXYGEN

NEEDED: Three candles, drinking glass, large jar.

EXPERIMENT 1: Light all the candles. Place the jar and glass over two of them. The flame will go out under the glass, and later will be extinguished under the jar.

EXPERIMENT 2: Try keeping insects or small animals in the jar after the flame has gone out.

REASON: The flame must have oxygen in order to burn. As the oxygen is depleted the flame goes out. The flame burns longer under the large jar because it contains more oxygen. It continues to burn when no jar or glass is placed over it because there is plenty of oxygen in the air.

One statement sometimes made in this connection is erroneous: the flame consumes all the oxygen. This is not true. It consumes a large part of the oxygen; after the flame goes out there is enough left under the glass to keep small animals or insects alive for quite a while.

Some fresh air is sucked into the space, too, as the air inside cools and contracts.

OXIDIZE IRON

NEEDED: Two tall jars with lids, a candle, steel wool.

EXPERIMENT: Wet a wad of steel wool and put it into one jar. Close the lids of both jars and let them stand two days. Light the candle, and let it down into the jar without the steel wool. It will burn a few minutes. Light it again, lower it into the jar with the steel wool, and its flame goes out quickly.

REASON: The steel wool rusts, and rusting is slow oxidation. The process uses up oxygen, so that after two days nearly all the oxygen in the jar is used up. There should not be enough to support the burning of the candle wick.

Perform this experiment where there is no breeze to blow the gases of oxidation out of the jar. Some steel wool is covered with an oil film to protect it from rusting in air on the dealer's shelf. To remove the oil, wash the steel wool in varsol then let it dry overnight to get the varsol off.

Oxidize Iron

MAKE INK

NEEDED: Fine steel wool, varsol or kerosene for cleaning it, white vinegar, tea bags, mucilage, containers.

EXPERIMENT: Clean a wad of steel wool with varsol and let it dry overnight. Put the steel wool into a jar, and cover it with

Make Ink

vinegar. Set it in a pan of water, hot but not boiling. Put four tea bags into half a cup of water, and boil. Let both solutions cool, mix them in equal amounts, and stir. Dip a finger into the mixture, and mark a large X on a newspaper page. It will gradually show up black.

REASON: Chemical action of the vinegar and iron produces hydrogen and iron acetate. The tea yields tannin. Mixed, they produce ferrous tannate, which is almost colorless but which changes to ferric tannate when dry and exposed to the air. In from three hours to a day the color change in the X should be complete—it should be black. A little mucilage mixed with the newly-produced ink will allow it to flow from a pen as regular ink.

ELECTROLYTES

NEEDED: A battery of three to six volts, salt, sugar, water, wire, glass containers.

EXPERIMENT: Connect the wires to the battery, bare the other ends of them, and hold the bare ends in water. Nothing happens. Put sugar in the water; nothing happens. Get fresh water, put salt in it, place the bare wires in it, and bubbles form on one wire.

REASON: Salt is a compound of sodium and chlorine. When the salt dissolves some sodium atoms give up electrons to become positive in charge; some chlorine atoms have grabbed those extra electrons and have become negative in charge. These charged atoms are called "ions."

The positively charged ions move to the negatively charged wire where they pick up electrons enough to make them neutral. But sodium metal reacts very rapidly and vigorously with water to give sodium hydroxide, Na(OH), and hydrogen gas. It is the hydrogen gas that is seen.

Pure water and sugar water do not form ions. Those that do and can carry electricity are called "electrolytes." Acids, bases, and salts form ions in many solutions.

Electrolytes

SOLUTIONS

NEEDED: Observation.

EXPERIMENT: Note that sugar dissolves more easily in hot tea than in iced tea. Note that a cold soft drink is full of gas, but when it gets warm it gets "flat" or loses much of its gas.

COMMENT: Solutions composed of ordinary solvents such as water can usually hold more solid solutes, such as sugar or salt, when they are warm. The opposite is true if our solutes are gases. Solvents in general can hold more gases in solution when they are cold.

Solutions

REASON: When a solid is dissolved in a liquid, the physical state of the solid may change to liquid, absorbing heat and lowering the temperature of the solution. Thus, solubility is increased as the temperature is raised.

In the case of gases in solution, raising the temperature increases the speed of the molecules of the gases and the solution, causing molecules of dissolved gas to leave the solvent.

Heating a solid is in the direction of melting it. Thus, it ought to "melt into" a liquid a littler easier.

Heating a liquid is in the direction of turning it to a gas. Thus, it ought to come out of the liquid a little easier, become a gas.

SOLUBILITY

NEEDED: Household iodine, water, mineral oil, a jar with a tightly-fitting lid.

EXPERIMENT: Into a jar half filled with water put a few drops of iodine, enough to make a light brown solution. Shake. Then put in two or three tablespoonfuls of mineral oil. Shake the mixture again, this time quite vigorously. Set the jar aside for a few minutes.

REASON: The color has left the water and has gone into the oil, showing that iodine is more soluble in mineral oil than in water.

(Vegetable oil does not work in this experiment.) This could be called solvent extraction.

Almost all substances are more soluble in some liquids than in others. The explanations of this can get quite technical and often controversial.

Solubility

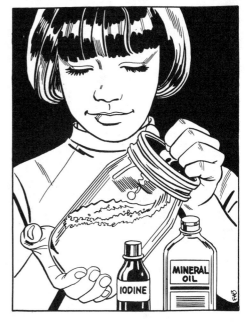

GROW CRYSTALS

NEEDED: Two ounces of alum from a grocery store, water, heat, string, a jar, a paper clip.

EXPERIMENT: Heat the alum in a half cupful of water until all the alum is dissolved. Pour the solution into the jar, and hang the paper clip in the solution. After a few hours crystals of alum will be seen growing on the clip.

REASON: When the alum is all dissolved we have a saturated solution, which means the water will not hold any more of the alum. As the water cools it cannot hold all that has been dissolved: it becomes a supersaturated solution and some of the alum leaves the water, forming itself into crystals. The "super" part crystallizes out.

If crystals begin to form in the bottom of the glass and not on the clip, touch the clip to some of the dry crystals then hang it back into the solution. Crystals should begin to form on it very soon. A book suggests trying this trick with table salt. The salt water solution is likely to start rusting the metals rather than forming crystals.

Grow Crystals

A COLOR MYSTERY

NEEDED: Water, corn starch, household iodine, a stew pan, heat.

EXPERIMENT: Fill a measuring cup with hot water, pour the water into the pan, and put a tablespoonful of starch into it. Boil a few minutes, stirring constantly. Let it cool, add a few drops of iodine, and a blue color results. Heat the mixture again, and the blue color disappears.

COMMENT: We give the name "starch iodine" to a certain union of molecules of iodine with soluble starch. At room temperature the mixture is bright blue. What happens to produce the blue color, and what happens to make the liquid become colorless when heated has not been clearly understood, but is becoming better understood as this book is written.

A Color Mystery

ICE AND SALT

NEEDED: Two containers, two thermometers, water, ice, salt.

EXPERIMENT: Put much ice and a little water in both containers. Place the thermometers in the containers and note that the temperature is the same in both. Pour salt into one container, and note that the temperature goes down.

REASON: Unless there is heat loss, the temperature of melting ice stays the same in plain water. But salt on the ice lowers its melting point, and its temperature goes lower than the normal 32 degrees. Almost any soluble impurity will lower the melting point of ice, but salt is the one used in home ice cream freezers.

Salt is the most common substance used to melt snow and ice on highways and streets.

An interesting sidelight is that Fahrenheit used a mixture of ice and hydrochloric acid to produce what he thought was the lowest temperature possible. He put it at zero degrees on his scale.

Ice and Salt

MAKE A SALT

NEEDED: Wine vinegar, baking soda, a tablespoon.

EXPERIMENT: Put vinegar into the spoon, then sprinkle soda into the spoon. Bubbles of carbon dioxide, CO_2, will form. Use barely enough soda to stop the bubbling. You have used an acid, the vinegar, and a base, the soda, and produced a chemical reaction which produced a salt.

Taste it! Taste the vinegar and soda before performing the experiment; the vinegar is very sour, the soda is bitter. The salt formed is not bad, its taste is near neutral.

COMMENT: Any kind of vinegar may be used, but wine vinegar has a more pleasant taste. Any acid will react with any soluble base to form water and a salt.

COVER THE SCRATCHES

NEEDED: Two tin cans, a knife, a cloth, water.

EXPERIMENT: Remove the labels from the cans. Make scratches on one can; leave the other unscratched. Wrap damp cloths around both cans. In a few days rust will have formed in the scratches.

REASON: "Tin" cans are iron, with a thin coating of another metal, perhaps mostly tin, which does not rust. Scratching exposes

Make a Salt

the iron, which will oxidize in the presence of air and water. Aluminum cans cannot be used in this experiment.

We now see why it is good to "touch up" or cover the scratches on an automobile or bicycle with paint. The scratches expose the iron to moisture and air and invite rust which can spread under the paint.

Cover the Scratches

SULFUR IN YOUR EGG

NEEDED: A boiled egg, an old silver plated spoon, a stainless steel spoon.

EXPERIMENT: Place some egg white on one end of each spoon, and some egg yellow on the other ends (or other spoons). Leave overnight. Next morning the silver plated spoon will have a dark spot where the egg touched; the stainless steel spoon will be unchanged.

REASON: Eggs contain sulfur which unites with the silver to produce silver sulphide, a black tarnish. The sulfur does not readily unite with any chemical contained in the stainless steel spoon.

Try mustard instead of egg. It, too, contains sulfur in sufficient quantity to tarnish silver. The tarnish can be removed, but it is recommended that old spoons be used in this experiment. Dimes and quarters dated earlier than 1965 contained enough silver to use for this experiment.

AN EGGSPERIMENT

NEEDED: Vinegar, a jar, a raw unshelled egg.

EXPERIMENT: Place the egg in the jar, and fill the jar with vinegar. Watch. Bubbles begin to form on the egg, and it will rise to the top. Perhaps it will go down and up again. Leave overnight and it will be larger and very soft.

REASON: Most of the hard part of the eggshell is calcium carbonate. The acid of the vinegar attacks and combines with the calcium carbonate to form carbon dioxide gas which adheres to the egg in bubbles. The attached bubbles make the egg lighter, and allow it to float to the top. If it loses enough bubbles there, it sinks again.

By morning the hard part of the shell will have been dissolved, making the egg soft. By osmosis, some of the vinegar is taken into the egg through the softened shell, increasing the size of the egg.

PHYSICAL CHANGE OF A CHEMICAL

NEEDED: Sulfur in an old spoon and an alcohol lamp or hot plate with which to heat it.

EXPERIMENT: Heat the sulfur over the flame or the hotplate, and note the changes. It first melts into a watery, straw colored liquid. With more heat this shifts to orange and then to red. Then it becomes a slow flowing deep brown color, then almost solid, and finally becomes liquid again, and boils with a yellow vapor at 444 degrees centigrade.

REASON: The color of an element such as sulfur depends on its temperature and also on the size of the unit mass of the element, on the number of atoms in the molecule.

Physical Change of a Chemical

SIMPLE CHEMICAL CHANGE

NEEDED: Sugar, an old spoon, and a stove.

EXPERIMENT: Place sugar in the spoon, and heat it until the sugar has become a black charred mass.

REASON: Sugar is composed of carbon, oxygen, and hydrogen. The heat changes the sugar to carbon, which remains in the spoon, and water, which leaves the spoon as steam or vapor. The chemical change that has taken place is written:

$C_{12}H_{22}O_{11} \quad 12C + 11H_2O$.

CRYSTAL PALACES

NEEDED: Photographer's hypo, a pan, a stove, some water.

EXPERIMENT: Place the hypo in the pan, pour in a small amount of water, and heat on the stove. Stir constantly. When the hypo is dissolved, set the pan aside where it will not be disturbed.

As the water cools and evaporates, the hypo forms beautiful crystals, many of them in a form that suggests palaces.

Each salt has its own particular crystalline form. Try table salt in this experiment and see the difference.

SUPERSATURATION

NEEDED: Photographer's hypo (sodium thiosulphate), water, a measuring spoon, a small pan, a pane of glass or a mirror.

EXPERIMENT: Put five small spoonfuls (level full) of hypo into the pan, then add one spoonful of water. Heat gently until all the hypo is dissolved. Let the liquid cool almost to room temperature, then pour some out on the glass so that it covers the glass. Let it cool some more. Then drop one crystal of hypo into the center. Crystals will form into a beautiful pattern.

REASON: The liquid when cool is supersaturated, that is, it contains more of the dissolved hypo than it normally can hold. The forces in it are very delicately balanced, so that an addition of another crystal, or perhaps a jarring will start the crystallization process.

CATHODIC PROTECTION (1)

NEEDED: Three jars, three large nails, sandpaper or emery cloth, galvanized sheet metal from a tinsmith, tin can metal, water, salt.

EXPERIMENT: Clean the nails with sandpaper or emery cloth. Drive one through the sheet metal and one through the tin can

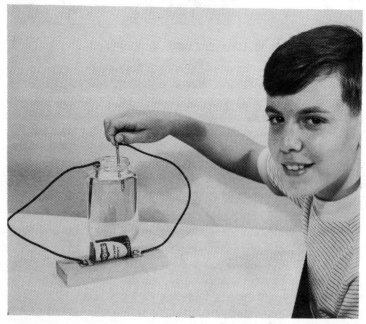

Cathodic Protection (1)

metal. Place one in each jar (one nail without other metal, as a control) and cover with water. After several hours two of the nails will be rusted; the one in galvanized metal will resist rust, perhaps for days.

REASON: The zinc coating on the sheet metal, in electrical contact with the iron of the nail, protects the iron because it is more active and corrodes first. It is called a "sacrificial anode." The tin can metal, if enameled, has no such effect. If the tin can metal is coated with tin it actually speeds up the rusting of the nail.

The practice of using a sacrificial anode to protect the hulls of iron ships is called cathodic protection. Most of the sacrificial anodes for ships are aluminum. Magnesium rods are used for the purpose in water heaters.

Cathodic Protection (2)

CATHODIC PROTECTION (2)

NEEDED: Two nails, salt water, a flashlight cell, connecting wire.

EXPERIMENT: Connect the cell to the nails and place the nails in the salt water so they do not touch. One nail will rust quickly, the other will not.

REASON: One terminal of the cell produces an excess of electrons, and the wire carries them to one of the nails, giving it a minus (−) charge. That nail thus attracts hydrogen (+) from the water. The hydrogen coats the nail, protecting it from oxidation (rusting). The other wire from the cell draws some electrons from the other nail, giving it a positive (+) charge attracts chlorine and other anions, forming oxidizing substances which attack the nail. Inside the cell one electrode is corroding and the other is not. The corroding electrode is connected by the wire to the protected nail.

This experiment illustrates control of the electrolytic corrosion of underground metal such as water pipes by the application of a negative electric current.

BLEACH

NEEDED: A glass half full of water, a few drops of red ink or food color, a teaspoonful of bleach.

EXPERIMENT: Mix the color and water, then add the bleach. The color disappears, leaving the water clear.

REASON: The bleach was made by reacting sodium hydroxide (lye) with chlorine to form unstable sodium hypochlorite, NaClO, which decomposes to liberate an abundance of nascent oxygen which is very reactive chemically. The oxygen atoms react with the color molecules to form colorless molecules.

Nascent, or "newborn" oxygen atoms are atoms that have not had time to combine into the oxygen molecule O_2. They are unattached, ready to attach themselves to other atoms or molecules. If they are combined into oxygen molecules they are not so reactive.

ANOTHER BLEACH

NEEDED: A glass jar, a small tin lid with a wire attached, some sulfur, matches, and a flower or apple peelings.

EXPERIMENT: Light the sulfur in the lid and let it down into the jar by means of the wire. Place the lid on the jar. After the flame has gone out, remove the dish of sulfur and place the flower or apple peels in the jar, covering it again quickly so that the gases do not escape. The flower or apple peelings will be seen to lose at least part of their color.

REASON: SO_2 (sulfur dioxide) gas formed by the burning S (sulfur) in the air will bleach many vegetable colors. It reacts with water to form sulfurous acid. The acid unites with the colored compounds in some organic substances and changes them to colorless compounds. Dried fruits have been home bleached and sterilized by this general process for many generations. A big tight barrel is generally used in "sulfuring" apples.

Another Bleach

PRECIPITATION

NEEDED: Epsom salts, ammonia solution, water.

EXPERIMENT: Make a solution of Epsom salts by dissolving some in water. Add a little ammonia solution. A white precipitate will be formed.

REASON: Epsom salts is magnesium sulfate. It reacts with the ammonia to form magnesium hydroxide, which is not soluble in water. It "precipitates" out in the form of white solid particles.

The chemical language for this is magnesium sulfate + ammonium hydorxide ammonium sulfate + magnesium hydroxide (precipitate).

EASY OXYGEN

NEEDED: A tall glass, cardboard, peroxide hair bleach from the drug store, baking soda.

EXPERIMENT: Put an inch of bleach into the glass, add a teaspoonful of baking soda, place the card over the glass to keep out the air, and let it sit an hour. Bubbles coming out of the mixture are oxygen. To prove it, place a glowing splinter of wood in the glass above the liquid and it will burst into flame.

REASON: The hydrogen peroxide molecules slowly change into plain water and nascent oxygen atoms. The O (oxygen) atoms

Precipitation

unite into pairs to form oxygen molecules, O_2, which move into the air space and accumulate.

Be sure the top of the glass is well covered by a smooth card so the oxygen will not escape. Put the splint into the air space as quickly as possible after removing the card.

PAPER CHROMATOGRAPHY

NEEDED: An old dish, paper towel, ink.

EXPERIMENT: Place a piece of paper towel on the dish, put a drop of ink at a corner of it, and watch. The ink will be of varying colors as it is soaked up through the paper fibres.

REASON: Most inks are made up of different substances, and those travel through the paper at different speeds. Brown ink is very good for this; some inks may not separate at all. But this principle is dependable enough for its use in chemical analyses. It is called "paper chromatography."

Place a quarter teaspoonful of ink in the center of a paper towel, and as it travels through the paper fibres, place a half teaspoonful of water in the center of the circle. This is shown in the photograph.

INK ERADICATOR

NEEDED: Lemon juice and a household laundry bleach.

EXPERIMENT: Put a spoonful of water into a cup and add a spoonful of bleach. With a toothpick, place some juice on the ink letter to be eradicated. Let it stand a minute, blot it, then place bleach on it, again using a toothpick. The ink should disappear slowly. Blot again, and add a little more juice to prevent damage to the paper. Blot again.

REASON: Citric acid in the lemon juice reacts with a chemical in the bleach to form an unstable hypochlorous acid solution. This solution decomposes to liberate uncombined "nascent" oxygen atoms which unite with the dye which colors the ink. The dye molecules are destroyed, leaving a colorless residue.

SULFUR DIOXIDE

NEEDED: Small can, lemon, photographer's hypo, heat.

EXPERIMENT: Put a few crystals of hypo into the can with a little lemon juice and heat the mixture. The strong choking odor that comes off is sulfur dioxide.

REASON: The chemical action that takes place between the photographer's hypo and the citric acid in the juice is interesting. Hydrogen citrate in the lemon juice combines with sodium thiosulfate (hypo) to give thiosulfuric acid and sodium citrate. The thiosulfuric acid when heated changes to water, sulfur, and sulfur dioxide.

A "CARBON VOLCANO"

NEEDED: Sugar, baking soda, an old spoon or tin can lid, and heat.

EXPERIMENT: Mix thoroughly a teaspoon of sugar and a teaspoon of soda. Place them on the lid or old spoon and heat the mixture. The "volcano" will begin to erupt and the "lava" will rise.

REASON: The heat drives out some of the water from the sugar, leaving carbon. The heat also drives out carbon dioxide gas and water from the soda. The gas and steam make the carbon "rise" very much as carbon dioxide and water (steam) make bread rise.

One volume of carbon dioxide is produced to each 12 volumes of steam when heat decomposes the mixed solids, and the mixture becomes light and frothy. The black carbon makes the mixture black.

DISTILLED WOOD

NEEDED: A tin can with tightly fitting lid, a match, some wood shavings or sawdust, a nail and hammer, a heat source.

EXPERIMENT: Make a small hole in the lid of the can with the nail. Place the wood in the can, cover with the lid, and heat the can on the stove. Steam will come from the hole, then gradually a gas will begin to come out. It will be a different color. The gas may be lighted with a match and it will burn. When no more gas comes out, remove the can from the heat, let it cool, and the wood will be seen to have turned to charcoal.

REASON: Wood is made of cellulose, with minor impurities of several kinds, all of which are mostly carbon, hydrogen, and oxygen. As the temperature goes above the boiling point of water, the water is driven off as steam. Soon tars and gases begin to come out of the hole and these will burn. Charcoal, which is mostly carbon, is left. This is the way artificial cooking gas is made, except that coal is used instead of wood, and the tars are saved and converted into many useful chemicals.

CHEMISTRY IN THE TOASTER

NEEDED: White bread toast, household iodine, water, a glass and a dish.

EXPERIMENT: Mix a teaspoonful of iodine in half a glass of water, and pour it into the dish. Dip the edge of a strip of toast into the solution. The center will turn bluish purple while the toasted part remains unchanged.

REASON: Iodine is a test for starch. The untoasted part of the bread turns purple, showing it is mostly starch; the toasted part remains unchanged because its starch has been changed by heat into dextrin. Dextrin iodide is not bluish purple. Starch iodide is.

Toasting is a step toward digestion of the bread. This is why toast rather than plain bread is recommended for sick people.

A CHEMICAL TEST

NEEDED: Boric acid, denatured alcohol, food jar lid, match, stainless steel spoon.

EXPERIMENT: Put a few crystals of boric acid into the lid. Add a teaspoonful of denatured alcohol and stir with the handle of the stainless steel spoon. Light the mixture and the flame will be seen to have green edges. As the alcohol burns away the flame will be a very beautiful green color. This experiment should not be performed in bright light.

REASON: Boron in the boric acid, H₃Bo₃, is heated to incandescence by the heat as the alcohol burns. The H part of the burning acid gives an unnoticed pale blue. The boron gives a noticeable bright green flame. Other elements give different colors in the "flame test."

The Flame Test

THE FLAME TEST

NEEDED: An alcohol lamp or Bunsen burner, table salt, baking soda, small pliers, steel wool, a small copper wire.

EXPERIMENT: Hold the end of the copper wire in the flame; note the green color given to the flame. Hold a small piece of steel wool in the flame with the pliers and notice the color. Drop a little of the salt and soda into the flame and notice the color.

REASON: Each vaporized metal has its characteristic flame color, and many can be identified in this manner. The salt and soda show the same color because it comes from the sodium, which is one of the elements in each. Some single colors are so intense that they hide other colors which may be present. Certain types of compounds vaporize more readily than others in the same flame.

SOLVENTS

NEEDED: Carbonated soft drink, salt, a dime.

EXPERIMENT: Make sure the mouth of the bottle is wet. Drop a little salt into the drink, place the dime on the mouth of the bottle. It will move up and down, showing that gas is coming out of the bottle.

REASON: The water of the soft drink has carbon dioxide, a gas, dissolved in it. Salt water can hold less carbon dioxide in solution than plain water. When the added salt dissolves in the water some of the gas has to come out.

This does not mean that water cannot hold many different substances dissolved in it. It can. In the soft drink there are many dissolved substances in addition to the salt and the carbon dioxide. It just cannot hold as much carbon dioxide when salt is added.

Incidentally, if there are possible harmful chemicals in the drink, carbon dioxide is *not* one of them.

CANDLE FLAME CHEMISTRY

NEEDED: A burning candle, a medicine dropper, some matches.

EXPERIMENT: Remove the rubber from the dropper, and hold the large end of the glass in the candle flame. Keep trying, and you will find a way to hold it so that gases coming through the dropper may be lighted at the little end. (Careful—the glass gets hot.)

REASON: The liquid paraffin climbs up the wick by capillary action, and is heated there and turned into gases. Ordinarily these would be burned in the outer portions of the flame, but they may go up the cool glass tube and be lighted at the upper end.

Much complicated chemistry goes on in the burning candle. Some of the substances found there are vaporized paraffin, ethylene, carbon dioxide, carbon monoxide, water vapor, hydrogen, oxygen, and nitrogen (from the air). The end products of the burning are mostly carbon dioxide and water.

LEAD CHEMISTRY

NEEDED: Inexpensive lead acetate from the drug store, zinc from a flashlight battery, distilled water, jars, soap and water, steel wool.

EXPERIMENT: Stir half an ounce of lead acetate in a pint of distilled water. Some tap water might do, if it is comparatively free of minerals. Let the solution sit overnight, and pour off the clear liquid for use in the experiment.

Scrub the zinc with soap and water, then with steel wool until it is bright. Suspend the zinc in the solution. Immediately it begins to turn black, as zinc replaces the lead in the solution. The lead deposit continues to grow on the zinc, and glistening spots may be seen. Leave it overnight. The zinc will be practically gone by next day. Take it out of the solution, let it dry a day or two, and it will be covered with bright shining lead crystals.

REASON: Zinc is more active than lead, which means that zinc atoms readily lose electrons if some units which absorb electrons are at hand. Lead in the acetate solution is in the form of lead ions which can attract and hold two electrons per ion. Zinc atoms lose two electrons per atom and become soluble zinc ions. Lead ions become insoluble lead atoms which unite to form glistening crystals.

Note: Lead acetate (common name sugar of lead) is poisonous if swallowed.

MORTAR AND CONCRETE

NEEDED: Lime, cement, sand, water. A builder will probably give a little lime, cement, and sand from a building site.

EXPERIMENT: Wash the sand in a jar or can to remove the dirt. Mix one spoonful of lime with water so that it makes a paste. Put in four spoonfuls of sand and mix thoroughly. Let it dry slowly on a metal lid. This is mortar.

Mix three spoonfuls of sand and one of cement with enough water to make a paste, and let it dry slowly. This is concrete. The drying should take place under a moist cloth to prevent a too-rapid loss of water.

REASON: As water evaporates from the mortar mix, carbon dioxide from the air takes its place, forming calcium carbonate, which is the chief part of natural limestone. Concrete dries by combining with water to form a hard firm complex. It must dry slowly, and it gets harder as it ages. Several years may elapse before it reaches maximum hardness.

The words "cement" and "concrete" are used sometimes to mean the same thing: the finished concrete. This is incorrect. "Cement" is highly active powder before it is mixed with sand and water. It is made by burning together limestone and clay or shale, then powdering the clinker. Just how it hardens into the popular building material so widely used is not adequately explained.

SUBSTANCES THAT SUBLIME

NEEDED: A jar with a loosely-fitting lid, a pan, some mothballs, water, a stove.

EXPERIMENT: Place the mothballs in the jar, cover with the lid (do not tighten the lid!), place the jar in the water in the pan, and heat the water until it is ready to boil. Turn off the heat and watch. Notice that the mothballs do not melt and become a liquid, but turn directly into a gas. They "sublime." Some of the gas condenses on the sides of the jar, making crystals. If the gas is allowed to flow out of the jar the mothballs eventually disappear, more rapidly if they are again heated.

REASON: Most substances melt and pass through the liquid state before changing into a gas. Water does usually, but clothes on a line will "freeze-dry"—that is, will dry when the weather is too cold to let the water in the wet clothes become a liquid.

CHANGING CHOCOLATE

NEEDED: Two chocolate bars.

EXPERIMENT: Let one of the bars get warm in the sun or elsewhere. Allow it to cool again. It will have become yellowish-white in color. Compare with the other.

REASON: According to Hershey's L.F. Santangelo, chocolate softens at 80 degrees, and melts at 92 degrees. At this temperature some of the cocoa butter separates from the other ingredients, then when the bar hardens again, the cocoa butter crystallizes on the surface as yellowish white crystals. This perhaps makes the chocolate less appetizing, but does not affect the nutritional value or the taste.

THE KITCHEN MATCH

NEEDED: A kitchen match.

EXPERIMENT: Strike the match. Note carefully how it burns when held horizontally.

REASON: When the head is rubbed a little heat is produced. Not much, but enough to light the tip, which is probably phosphorus sulphide and potassium chlorate and a binder. The burning tip lights the bulk of the head of the match, which is probably potassium nitrate and charcoal with a binder. The burning head lights paraffin in which the end of the match stick has been dipped, then the burning paraffin lights the wood which is probably pine.

The other end of the stick (that end held in the fingers) had been moderately "fireproofed" by dipping in a chemical (probably ammonium phosphate) intended to actually retard the flame, to prevent the stick from glowing when discarded.

THE SAFETY MATCH

NEEDED: A match book.

EXPERIMENT: Strike a safety match on the part of the book intended for the purpose. Also, rub the match quickly over a glass surface and see if it will strike.

REASON: The head of the safety match is composed of antimony sulfide and potassium chlorate, with a binder to hold them together. This mixture does not ignite easily by the heat of friction, but is lighted by any tiny flame.

The "strikum" on the paper book is composed of red phosphorus, powdered glass, and glue—the binder. As the head of the

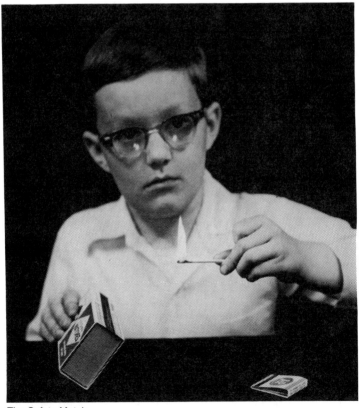

The Safety Match

match is rubbed across the "strikum" a little of the phosphorus is torn off and takes fire in contact with the very combustible head of the match. The head takes fire and lights the paraffin soaked stick. A safety match may be struck on glass because the smooth surface can allow a build-up of heat by friction—enough to fire off the head before the head wears out, as it would do on most surfaces.

Match companies do not all use the same ingredients in their matches. Some other chemicals used are red lead, lead dioxide, and manganese dioxide. White phosphorus is no longer used because it takes fire too easily and is poisonous.

There are some safety matches difficult, if not impossible, to strike on glass, because of the difference in ingredients used. Also, it takes a little practice. Motion must be rapid and pressure just right.

Chapter 14
Mechanics

THE RUBBER ERASER

NEEDED: A pencil eraser, a rubber balloon, pencil, paper.

EXPERIMENT: Notice how the pencil eraser will erase the marks, while the rubber balloon has more of a tendency to smear them.

REASON: The pencil eraser has tiny pieces of hardened gritty rubber held in position by the rubber film. As the eraser is moved, the small gritty pieces remove the graphite marks by removing the upper surface of the felted paper. The balloon has only a smooth rubber film which spreads the marks over the surface of the paper.

A JET-PROPELLED BOAT

NEEDED: A piece of wood, a medicine dropper, a balloon, a place to float it.

EXPERIMENT 1: Make the boat as shown. It will be propelled through the water as the air leaves the balloon through the dropper.

REASON: When the balloon is blown up, the stretched rubber pushes on the air inside it. When there is an opening, as through the dropper, the air is pushed out by the balloon and the balloon is pushed by the air causing the boat attached to the balloon to move.

Newton's Third Law: "Every action has an equal and opposite reaction."

EXPERIMENT 2: Sail the boat so that the end of the dropper is below the water surface, then again when the opening is above. The boat is likely to go farther when the opening is below the water, not because it pushes against the water more than against the air, but because the air leaves the balloon faster when allowed to exhaust above the water. At the faster speed, the wind and water resistance is greater for the boat.

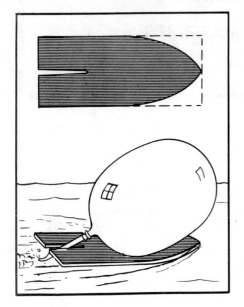

A Jet Propelled Boat

MUSCLES AND LEVERAGE

NEEDED: A broom and three people.

EXPERIMENT: Have two people hold the broom tightly as shown, so that the end is a foot or more above the floor. Ask them to bring down the broom handle so that it will come quickly into the circle or ring on the floor. You can push the broom aside with one finger so that it will not hit the ring.

REASON: Because of the leverage, a small force exerted by the finger at the end of the broomstick can easily change the direction of motion of the stick. Coordination of the two men has something to do with it also. The two people trying to bring the stick into the ring cannot work together; actually more often, they will find they are pushing against each other.

The greater length of the stick to the child's hands allows him to produce greater torque than the men, who have a length of stick much closer to the center of rotation.

Muscles and Leverage

THE CANDLE SEE-SAW

NEEDED: A candle, a long needle, matches, a newspaper to catch the wax.

EXPERIMENT: Push the needle through the candle, then light it at both ends. It will "see-saw."

REASON: The wax will melt faster on the lower end because the flame from the wick will heat it more. As that end gets lighter, the other end will come down and start to melt faster. The process is repeated, making the candle see-saw slowly.

The Candle See-Saw

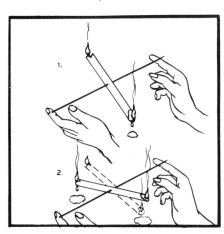

YOU AND A HORSE

NEEDED: A ruler, a scratch pad, a pencil.

EXPERIMENT: Measure the height of the stairs or steps, then find out how much energy is used in climbing them.

METHOD: Multiply the height by your weight to get foot-pounds of energy. Suppose you weigh 110 pounds and the stair is ten feet high, you have used 1110 foot-pounds of energy. To change this into horsepower, another figure must be added: time. Horsepower is 33,000 foot-pounds per minute, or 550 foot-pounds per second. If you climb the stairs in five seconds, then your power is 1100 divided by five times 550, or four-tenths of a horsepower.

DEFYING GRAVITY

NEEDED: A hammer, a ruler, a string, a table edge.

EXPERIMENT: Tie the ruler and hammer together as shown, and they will hang from the table in what will look like a most precarious manner.

REASON: Most of the weight is in the hammer head, and if the ruler is moved along the table edge until the center of weight of the assembly is directly under the edge, the balance point will be easy to find. It is as if the weight were hanging straight down from the table. The center of gravity of the assembly must be on the table side of the table edge.

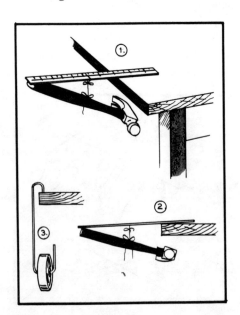

Defying Gravity

MULTIPLIED MUSCLE POWER

NEEDED: A hammer and a nail, a piece of wood, a small block of wood.

EXPERIMENT: Try pulling the nail with the nose of the hammer against the wood. Then place the small block under the hammer as shown, and the nail will be pulled easily.

REASON: Note the two broken lines in the drawing. If the distance of line A is one inch, and the distance of line B is 10 inches, the pull on the handle places about ten times as much pull on the nail. If 40 pounds of pull is exerted on the handle, about 400 pounds of pull is exerted on the nail. The pulling force applied to the handle will move ten times as far as the nail moves.

The hammer is a form of lever.

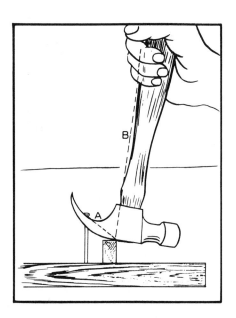

Multiplied Muscle Power

WEIGHT LIFTING

NEEDED: A weight.

EXPERIMENT: Lift the weight as in the drawing at left, and it is easy. Try to lift it as shown in the drawing at right, and it is difficult or impossible.

REASON: In the left drawing, the muscle tension (force upward) and the weight act along the same line and are equal. Both are fairly small.

In the right drawing, where the arm is extended, the muscle tension times its distance from the pivot point must equal the weight times its distance from the pivot point (length of arm). Since the weight is far from the pivot point and the muscle close to the pivot point, the muscle tension must be many times the weight, if it is to support the weight.

The lower drawings show this in an over-simplified manner. The pivot point is the shoulder joint.

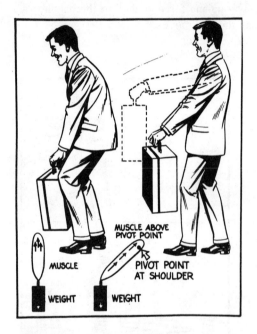

Weight Lifting

GLAMORIZING THE WEDGE

DICTIONARY DEFINITION: A wedge is a piece of wood or metal, small at one end and larger at the other, used for rending or compressing.

COMMENT: A wedge is a type of inclined plane which is pushed into an object to cut or split it. The smaller the angle of the wedge, the easier it is to cut the object; therefore a sharp knife cuts better than a dull one.

The push required to move a wedge into an object is not easy to determine because of friction.

The wedge is used by carpenters and woodsmen in the form of the ax, chisel, plane, and nail. The farmer turns his soil with a wedge—the plow. A rotating wedge or cam is used to push up the valve rods in automobile engines. A needle is a wedge, too.

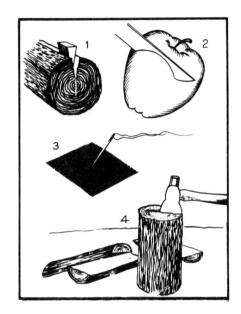

Glamorizing the Wedge

TRANSVERSE WAVE MOTION

NEEDED: Hose or rope, a large pan of water, a cork, small stone.

EXPERIMENT 1: Move the hose or rope up and down to make transverse waves as shown in the drawing.

EXPERIMENT 2: Drop the stone into the water. The floating cork does not move outward with the waves, but moves up and down.

REASON: The wave form moves down the length of the hose as the hose itself moves only up and down.

Similarly, the wave form from the disturbance in the water moves over the surface of the water as the water particles move up and down. The cork shows that this is true, as it moves up and down.

THE SQUEEZE BOTTLE

NEEDED: A flat-sided bottle filled with water and a thin card or paper to cover the bottle mouth.

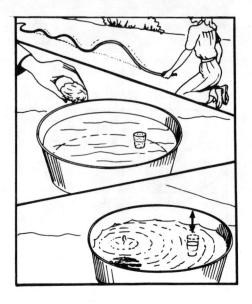

Transverse Wave Motion

EXPERIMENT: Place the paper over the bottle mouth, invert the bottle, and the water will not run out because atmospheric pressure holds it in. This is shown in A. Squeeze the bottle on the flat sides and some of the water will come out.

REASON: The bottle, while made of glass, actually is elastic, and may be distorted by squeezing so that some of the water is forced out. If it is held long enough for warmth from the hands to go through the glass, the expansion of the water due to the heat will force a little more out.

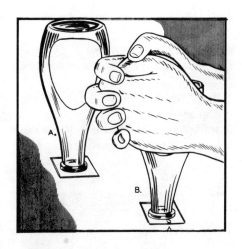

The Squeeze Bottle

BUOYANCY

NEEDED: A rubber balloon and a pan of water.

EXPERIMENT: Blow up the balloon, and see how difficult it is to push it under the water.

REASON: The water in the pan is much heavier than the balloon and the air in the balloon. The force required to move the balloon down into the water is equal to the weight of the water raised or "displaced" by the balloon. Archimedes discovered this principle more than 2200 years ago.

If we wanted to be strictly accurate, we would have to say "the additional force required," because the weight of the balloon, although slight, would displace a small amount of water.

ELASTICITY

NEEDED: A rubber balloon.

EXPERIMENT: Inflate the balloon with the breath. It is difficult or impossible with some new balloons. Now stretch the balloon several times with the fingers. It is easier to inflate it with the breath now.

REASON: Rubber, while usually thought of as very elastic, is not one of the most elastic substances. When the balloon is stretched

Elasticity

with the hands it never goes back to the tough, tight, difficult state it was in before stretching.

A definition of "elastic" is: "Able to return immediately to its original size or shape, after being altered by squeezing, stretching, compressing, or bending." Rubber does not fit that definition. Rubber is very "stretchable," however, and this is one of the characteristics of elasticity.

MORE ELASTICITY

NEEDED: Rubber band, strips of scrap glass, a vise with plastic jaw inserts, gloves, protective goggles.

EXPERIMENT: Clamp the glass in the vise rather tightly. The plastic will help prevent it from cracking. Wiggle one of the glass strips back and forth. It aways returns to its original shape.

Now measure the length of the rubber band, and stretch it as far as possible several times. Measure it again. It is longer than before.

A definition of "elastic" is: "Able to return immediately to its original size or shape, after being altered by squeezing, stretching, compressing, bending, etc." Glass is more elastic than rubber.

Get two long strips of scrap glass from a hardware store. Place on a book or two, as shown, and press down on the center of one. It will bend a surprising amount before breaking.

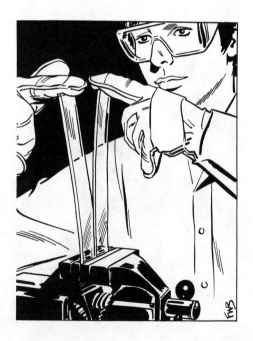

More Elasticity

Friction of Water

FRICTION OF WATER

NEEDED: Two similar jars with tight lids, one half filled with water, and a slightly inclined board.

EXPERIMENT 1: Release the jars at the same time at the top of the incline. The one containing water will roll faster at first, but the other should roll farther when it reaches the level of the floor.

EXPERIMENT 2: Try this by filling one jar half full of water and filling the other completely with water. Which will roll farther?

EXPERIMENT 3: Fill one jar completely with water; leave the other without water. See which will roll farther.

REASON: Friction between the water and the sides of the glass jar makes the jar slow down. Air in the "empty" jar produces no such friction.

ROLLING RIGHT ALONG

NEEDED: Two cans with lids and sand.

Rolling Right Along

EXPERIMENT 1: Fill the jars with sand. Stand one on end and pull it along on the table or floor. Roll the other. Note that the can on end is much harder to move along.

EXPERIMENT 2: Pour out half the sand in one jar, and try to roll it. It will not roll smoothly at all.

REASON: When the can of sand is pulled along, dragging on the table, small imperfections in the touching surfaces are always catching on one another. The friction produced must be overcome before the can can move. When the can is rolled it travels over the bumps rather than through them.

When a can is half empty, rolling the can part way builds up the sand on one side. If released the can will roll backward until the sand is balanced on both sides of the can. Continuous rolling involves continual lifting of part of the sand. This friction is between the sand and the side of the can, and between sand and sand.

THE BRAKE

NEEDED: Cardboard, a pencil, something for marking a circle.

EXPERIMENT: Cut a circle from the card and push the pencil through the center. Twirl the newly made "top" on the table. It will turn for several seconds. But if the pencil is touched between finger and thumb it stops quickly.

REASON: The friction between the pencil and finger slows the top down. If not touched the friction between the pencil and the table and between the card and the air will gradually stop the motion.

The fingers illustrate the principle of the brake in the automobile. When the brake pedal is pushed a material is pushed against a turning wheel or drum, causing friction which tends to slow the motion.

The Brake

THE WHEEL

NEEDED: Meter stick or yardstick, toy wagon with rubber tires, a flashlight battery, foot rule.

EXPERIMENT: Place a marker on the ground exactly under the rear axle of the wagon. Place the end of the meter stick exactly over the axle, on the tire. Hold the stick on the rubber as the wagon moves along, letting the stick move forward without slipping as the wheel turns. Stop when the other end is exactly over the axle.

Measure the distance from the marker to the axle. Guess: will it be the length of the stick? Half the length of the stick? Or twice the length of the stick?

REASON: The top of the wheel moves twice as fast as the axle when the wagon is pulled along. The point of the tire touching the ground does not move at all—at the instant it is in contact with the ground.

This may be tried on a table top with a rule and round object. The pencil shown is the marker.

The Wheel

THE BRIDGE (1)

NEEDED: Strips of cardboard about 4 × 12 inches, books, jars, sand.

EXPERIMENT: Make a bridge by putting two strips of card between stacks of books as shown in the upper drawing. It will not hold up an empty jar. Make the bridge as shown in the lower drawing, place the jar on it, pour sand into the jar. The bridge is stronger.

REASON: Materials may be stressed several ways: compression or pushing together, tension or pulling apart, flexure which means bending, torsion or twisting, shearing or slicing apart.

The cards will bend easily when weight is put on them as in the first try. But when one card is placed below, it is pushed on with a compressionable force. It is quite strong when compressional or tensile forces are applied, and so holds up the upper card which rests on it.

The Bridge (1)

THE BRIDGE (2)

NEEDED: Regular cardboard, corrugated cardboard, books, weights.

EXPERIMENT: Place a strip of regular card about 2 × 15 inches across books to make a bridge. It will hold up little or no weight. Place a similar strip of corrugated board across, and it will hold up a weight such as a glass paper weight. Split the corrugated card, along its length as shown, and it will be weak, like the regular card.

REASON: The corrugations between the flat card form triangles that are very strong in resisting compressional and tensile forces. In other words, to bend the card the corrugations would have to be pushed or pulled with enough force to deform them.

When the card is split through the corrugations the sections of it will flex, or bend when only a small force is applied. There is practically no compression and tension to be resisted. The split card behaves like the regular card.

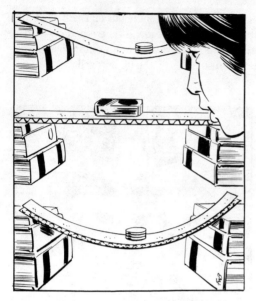

The Bridge (2)

FRICTION

NEEDED: Roller skates.

EXPERIMENT: Have the skates pointing forward, and try to walk. Moving forward is difficult because the skates offer little friction between the feet and the floor. This is rolling friction, and since the wheels are between the feet and the floor it is easy for the wheel to roll but not easy to move the body forward.

Without skates the shoes against the floor offer so much friction that they slide only with some difficulty.

COMMENT: Friction can produce heat and wear in bearings and be very destructive, but it is useful in almost all our actions. Even in sleep it is useful; it keeps us from sliding off the bed.

Without friction an automobile could not move and without friction it could not stop once it started moving.

THE COMPOUND PENDULUM

NEEDED: A paper cone (a drinking cup will do), two lengths of string, a doorway or other support, a large square of cardboard, preferably dark in color, some salt.

EXPERIMENT: Hang the strings from two places at the top of the doorway, tie the cup to each string as shown in the drawing. Make a hole 1/8-inch in diameter in the bottom of the cup.

Draw the cup to one corner of the cardboard, fill it almost full of dry salt and release it. As the cup swings this "compound pendulum"

Friction

will make various and interesting designs with the salt. Change the length of the upper string or the lower string, try again and the design will be different.

REASON: The strings and cup act as two different pendulums swinging back and forth through the doorway, also to and fro across the face of the doorway. This combined motion causes the cup to swing in a regular pattern giving the beautiful designs.

The Compound Pendulum

ANGULAR MOMENTUM

NEEDED: An object on the end of a string.

EXPERIMENT: Whirl the object around on the string. Extend a finger so that the string begins to wind up on it. The finger may be held still, yet the object will continue to whirl, faster and faster as the string is wound up on the finger, until finally it touches the finger.

REASON: The angular momentum is the mass of the object, times the angular velocity (that is, the rate of spin), times the length of the string squared. The angular velocity remains constant. Since the mass of the object is constant, the rate of spin must increase as the string gets shorter to keep the product unchanging. So, the object spins faster and faster.

FLUIDICS

NEEDED: A steady flow of water from the tap, a soda straw, a glass.

EXPERIMENT: Fill the mouth with water, then blow the water out of the mouth through the straw so that it hits the larger stream from the tap. The smaller stream from the straw will turn the larger stream aside.

REASON: This is the principle of the fluid amplifier device in a new technology called "fluidics." The weaker stream of water

Fluidics

changes the larger stream, much as a weaker current or charge can change a larger current in electronics. Fluidic devices are replacing electronic devices in many control systems today. They have advantages in some cases.

HAMMER—HEAD

NEEDED: A nail, a hammer, a plank, a large book such as a heavy dictionary.

EXPERIMENT: Place the book on someone's head, place the plank on the book, and the nail may be driven into the board without hurting the head.

REASON: There is much unfilled space between each page of the book and this space contains air which acts as a cushion. The mass and weight of the book absorb much of the blow of the hammer, because of their inertia. The book also spreads the force of the blow over a wide area of the head. If the book used is small, a tack hammer and a tack should be used instead of a hammer and nail.

STICK-SLIP FRICTION

NEEDED: A violin or a rosined string.

EXPERIMENT: Draw the bow along the violin string, and notice that the string vibrates, or moves rapidly back and forth.

Stick-Slip Friciton

REASON: As the bow moves, the string sticks to it and moves with it. Then the string slips and flies back. The bow sticks to it again, and moves it again. The process is repeated to make the musical note.

This type of friction may be shown by drawing a rosined string along the edge of a tin can or through a hole in the bottom of a can, or the string can be attached to the bottom of a can, stretched, and rubbed with a pencil. An unpleasant growling sound is made in this way, whereas a musical note is produced by the tuned violin string. Slip-stick friction may also be shown by rubbing the finger across a table as shown in the upper drawing.

THE HOMING CAN

NEEDED: A coffee can, rubber bands, a heavy iron bolt or nuts.

EXPERIMENT: Fasten the bands to the bottom and lid of the can, and attach the iron object in the middle of the bands. Put the lid on. Roll the can away and it will return.

REASON: As the can rolls away the hanging weight hangs and does not turn as the bands twist. This twisting builds up as energy in the bands. As the can stops rolling the energy can be released from the bands as they begin to untwist, rolling the can along in the opposite direction.

The Homing Can

THE MYSTERY SPOOL

NEEDED: A spool with thread wound around it.

EXPERIMENT: Place the spool on a level surface. Pull the thread as if to unwind it, and the spool will roll toward you winding up more thread.

REASON: The frictional force acting where the spool makes contact with the surface causes a greater torque about the center of the spool than does the force due to the pull on the string. The result is angular motion in a direction to wind up the string.

NEWTON'S THIRD LAW

NEEDED: A smooth board and sticks to hold a pendulum, large soda straws, string, large bolt or other weight.

EXPERIMENT: Build the apparatus as shown, so the board can move on the straws. As the pendulum swings the platform moves back and forth in the opposite direction.

REASON: Newton's third law states that for every action there is an opposite and equal reaction. At the end of a swing the string pulls one way on the weight and the other way on the platform, and so both move.

When the support is allowed to move in this way the pendulum behaves as if it were shorter than it really is. The "virtual pivot" is between the support and the bob (weight).

Newton's Third Law

The I-Beam Idea

THE I-BEAM IDEA

NEEDED: A yardstick and a weight.

EXPERIMENT: Hold the yardstick by the ends, with the weight suspended from the center. If the stick is on edge, it does not seem to bend. If it is flat, it bends under the weight.

REASON: Engineers have discovered that a support beam varies in stiffness as the cube of the vertical height, other factors being equal. In the upper picture the cube of the vertical height is many times greater than the cube of the vertical height in the lower picture. Steel I-beams have much greater stiffness if the I is vertical. Observe the construction arrangement of the steel frame of a new building. Fishing rods bend easily at the tip where the vertical thickness is very small.

PULLEYS

NEEDED: Pulleys, cord, a support, weights.

EXPERIMENT: String the pulleys in various ways, and see how the weight is easier to move as more strands of cord are used.

REASON: The actual weight of the object is felt as the cord is pulled downward over a single pulley. If the pulley is placed at the bottom and the cord pulled upward, the weight seems to be about half as much, but the cord has to move twice as far as the weight moves.

A block and tackle, which consists of several pulleys and several strands of rope increases mechanical advantage. But as the number of strands supporting the weight is increased the applied force must act over a longer distance. In the drawing the second pulley system has the greatest mechanical advantage where four strands support the weight. The third and last arrangements have the same mechanical advantage, since each has two cords supporting the weight. In any case if the pull is vertical the theoretical mechanical advantage equals the number of strands which support the weight. More work must be put into any machine than is gotten out of the machine. The extra work must be done because of friction.

Pulleys

FOUCAULT'S PENDULUM (1)

NEEDED: Heavy weight, strong string, a solid support more than ten feet high, a four-foot stick.

EXPERIMENT: Start the weight swinging in a straight path, and place the stick under that path, to mark it. After ten or more minutes the weight will have changed its path of swing slightly—a proof of the turning of the earth.

REASON: Jean Bernard Foucault was a French physicist (1819-1868). He made discoveries in light, electricity, and magnetism, and invented the gyroscope. His pendulum experiment, proving that the earth turns, was made in 1851.

This is not an easy experiment to do at home. For a weight, the author used a 21-pound iron ball. A water bucket full of wet sand might be used. The author attached a bracket to a large tree limb 19 feet above the ground, and allowed the nylon cord to swing from it. A breeze or the untwisting of the cord or possibly other factors may interfere.

FOUCAULT'S PENDULUM (2)

NEEDED: Phonograph, large glass jar, stick, string, weight.

EXPERIMENT: Set up the apparatus as shown. Start the weight swinging, then start the turntable at its slowest speed. Note that the plane of swing of the weight does not change as the jar turns.

RESULT: The Foucault pendulum continues to swing in the same plane as the earth turns under it. If a phonograph is not at hand, a revolving stool may be used. Be sure the weight (which can be a fishing sinker) is suspended from the center, otherwise centrifugal force as the turntable turns may throw the experiment out of balance.

Foucault's Pendulum (2)

TRICKY MARBLES

NEEDED: A ruler with a groove along its length, a dish, two marbles.

EXPERIMENT: Place a marble on the dish, hold the other in the groove on the ruler and aim the groove at the first marble. When the second marble is allowed to roll down and strike the first, it stops and the first marble is catapulted out of the dish.

REASON: The marbles are elastic. When one hits the other, its force is transmitted into the second. Since the collision of the marbles is almost perfectly elastic, the momentum of the first marble is transferred to the second marble. The first marble stops and the second marble moves off with the original velocity of the first. (Glass is more elastic than rubber.)

Tricky Marbles

WHICH FALLS FIRST?

NEEDED: A yardstick or meter stick, a coin such as a quarter or half dollar.

EXPERIMENT: Place the stick against the wall so that it will not slip. Place the coin on the end, as in the upper drawing, and when the stick is released the coin will not fall as rapidly as the stick. Place the coin back one-third the length of the stick, and both the coin and stick will fall together.

REASON: The stick is not falling as freely as the coin. The stick is rotating about an axis at the wall. The velocity of the end of the stick is greater than the falling coin. The velocity of a point on the stick one-third the length of the stick from the end moves at the same velocity as a freely falling body.

Which Falls First?

THE TOUGH BALLOON

NEEDED: A balloon.

EXPERIMENT: Blow air into the balloon, lay it on the table, then burst it by placing a smooth surface, such as a book, on it and pressing down. The effort required to burst the balloon will be surprising.

REASON: Pressure is force per unit area. Here the pressure difference on the inside and outside of a toy balloon necessary to burst it is not very great. By putting the book on top of the balloon the force (push) is distributed over such a large area of the balloon that the pressure increase is small. A small push on a pin would cause a large pressure which would easily break the balloon. Total force = pressure × surface area.

A BAD BALANCE

NEEDED: Ruler and string, coins.

EXPERIMENT: Suspend the ruler on the string above a table top, and try to balance coins on it. It does not work. Put a screw eye in it, tie the string to it, and the ruler becomes a balance that can be used to show the principle of the lever.

REASON: The ruler cannot balance if suspended by a string from the lower side, because the center of gravity is above the suspension point or fulcrum. Either one side or the other will go down and touch the table top. Suspending it by the screw eye places the balance point above the ruler, and a balance can be effected as long as the coins are not stacked too high.

A Bad Balance

METEORS IN THE HAND

NEEDED: Flint rock and a piece of steel, or a cigarette lighter.

EXPERIMENT: Strike the rock with the steel, or turn the wheel of the lighter. Sparks will fly.

REASON: When the steel is rubbed over the flint quickly, some tiny pieces of the flint are rubbed off. The combination of the friction and the high speed at which they leave the flint heats them to the burning temperature of dry tinder or lighter fluid.

This is something like the meteors in the sky. Their speed through the air causes friction which heats them to a high temperature so that they give off light.

CRACK OF THE WHIP

NEEDED: A whip that will "crack."

EXPERIMENT: Try to think of the explanation of the loud sound made by the whip as the wave goes to the end, faster and faster, finally ending with the "crack."

REASON: It could be called an example of conservation of angular momentum; the momentum starting at the thick part of the whip continues, but the smaller parts of the whip must move faster to maintain the momentum.

The end of the whip moves faster than the speed of sound; it "breaks the sound barrier," moving faster than a .45 pistol bullet! The whip does not slap back on itself to make the noise.

UPHILL ROLL

NEEDED: Two conical paper cups, two yardsticks, three books, gummed tape or needle and thread.

EXPERIMENT: Sew or tape the cups together so that they make double cones. Arrange the yardsticks so they are held by the

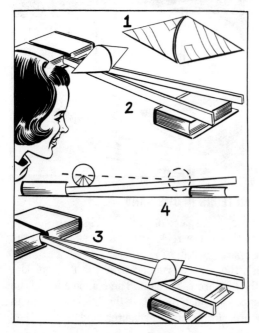

Uphill Roll

book in an upward slanting position as shown. The cups, placed at the bottom, roll to the top of the yardstick incline.

REASON: The cups do not actually roll uphill. Because of their shape, their center of gravity moves down as they roll to the upper ends of the spreading sticks. For proof, notice how far they are from the table at the beginning of their journey and at the end. They are closer to the table at the end.

THE ROLLING MARBLE

NEEDED: A ruler, a thick book, a marble, a rug on the floor.

EXPERIMENT: Place the end of the ruler on the edge of the book, let the marble roll down the ruler and measure the distance it rolls. Place the ruler so that the book is under the middle of it, let the marble roll from this middle point, and measure the distance it rolls. The distances rolled should be the same.

REASON: The original potential energy of the marble is the same in both positions mentioned above. Potential energy is the product of weight times height. This energy is converted to kinetic energy of motion in rolling down the ruler, therefore the speed of the marble is the same when it leaves the ruler in both cases and should roll the same distance along the rug. A very soft or deep pile rug is not too satisfactory for this experiment.

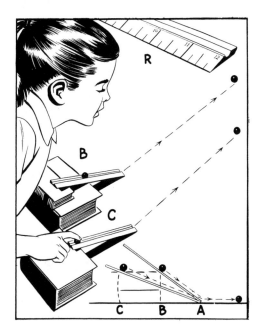

The Rolling Marble

More Rolling

MORE ROLLING

NEEDED: An incline, a roller skate, a ball, a cylinder, and a disc or wheel.

EXPERIMENT: Release all the rolling objects at the same time at the top of the incline, and see which reaches the bottom first.

REASON: All the objects have the same potential energy per pound at the start. As they move down the incline their potential energy is changed into kinetic energy, energy of motion, which is in two forms, translational and rotational. The translational energy is what gives a body a velocity straight ahead, while the rotational energy is tied up in making the mass of the body turn or spin. The skate has the greatest percentage of its energy as translational kinetic energy, the least as rotational kinetic energy, and so has a higher velocity all the way down.

The incline must be smooth, so that the effects of rolling friction need not be taken into consideration. See which object rolls farthest after leaving the incline.

THE SWING

NEEDED: A rope swing.

EXPERIMENT: Twist the rope, and let it start to unwind with your legs outstretched; the spin will be rather slow. Draw in your legs, and the spin will be faster. Extend your legs, and the spin will slow down again.

REASON: This is another study in conservation of angular momentum, which, for each particle in your body, is its mass times its angular velocity (rate of spin) times the square of its distance from the center about which you are turning. If you change the distance of some parts of your body from the center, the spin will have to adjust itself so that the angular momentum stays the same, (stick legs out and slow down, pull them in and speed up).

Skaters and ballet dancers use this principle. They start a spin or whirl with arms extended, then spin faster as they draw their arms in closer to their bodies.

Index

A

Acrobatic rice puffs	232
Adhesion, cohesion	171
Air, elastic	263
expansion	265
moisture, measure	42
polution	259
pouring cold	256
pressure trick	265
weight of	244
Alarm, power-off	51
Alcohol lamp, make your own	41
rub, why	290
Aluminum magic	231
An air horn	26
air can	266
eggsperiment	356
eye trick	135
electric pendulum	233
hole in the hand	70
illusion	107, 108
impossible trick	85
indoor greenhouse	108
old freezing question	274
optical illusion	79
Angular momentum	388
Another bleach	360
myth exploded	269
rubber band trick	285
straw through potato trick	90
Arms, the rising	103
Arm, valves in the	118
Arrow	256
Atomizer, soda straw	238

B

Bad breath	331
Balance, a bad	397
a hatchet	9
Balancing	125
Ball, an erratic	261
bounce the	262
the mischievous	251
the rising	84
Balloon, bashful	213
fallacy	269
in a jug	240
inside-out	244
musical	22
the dancing	249
the erratic	271
the obliging balloon	273
the stubborn	267
the tough	396
trick	240
Bands, mach	300
Barometer, a bad	270
needle	209
Bathroom, magnet in the	237
Beans, jumping	73
Bell assembly for detector	48
Bernoulli with a dime	95
Beverage, carbonated	268
Big sea, little sea	35
Bird in a cage	293
Bleach	360
flowers	341
Blinky the blinder	64
Blow bottle	252
Blow the coin over	96
Blowing snow	169
Board magic	228
Ouija	79
Boat, jet-propelled	369
of holes	164
Book, a cool	87
Bottle, a cucumber in	122
a mystery	246
a pop	338
bumping	11
squeeze	276
the squeeze	377
trick	242
Bounce a ball	262
Brake	382
Bread mold	104
Breath, holding the	126
Bridge (1)	384
Bridge (2)	385
Broken magnet	210
Brown cup-cakes	324
Bubble magic	226
rainbows in a	312
starchy	280
Bugs away	316
Build your own scale	43
Bulldog, the	25
Bullets, water	167
Bumping bottles	11
Buoyancy	195, 379
Burn flour	337
Burning paper	287
Buzzer	53

C

Cabbage, technicolor	335
Cage, a bird in	293
Can, an air	266
crushed easy	265
noisy	24
the homing	390
Candle, edible	92
flame chemistry	366
flame shadow	313
Candling, egg	119
Canon, smoke ring	272
Capillarity	165
capillary attraction	177
Capture spider webs	149
Carbonated beverage	268
Carbon dioxide and yeast	146
Carbon dioxide by fermentation	339
make	339
Case of the disappearing crystals	344
Cathodic protection (1)	358
protection (2)	359
Cavities, water	181
Center of gravity	9
Changing chocolate	368
Chemical action	341
physical change of a	356
& light	340
candle flame	366
in the toaster	364
lead	366
and curds	324
Chicken fountain	243
Chimes, knife & fork	17
Chocolate, changing	368
Chromatography, paper	362
Clean copper or brass	314
Clear ice cubes	185
Clinging water	178
Cloth & sponge mystery	159
fireproof a	286
Cloud in a jug	188
what holds up the	159
Coffee, tasteless	118
Colors, mystery	308
tricky	310
Compass cut-ups	209
homemade	220
Concrete & mortar	367
Cone, vibration through a	27
Contact angles, dry solids	183
or expand	289
Copper, salt & vinegar on	334
Cotton, fast	281
Cover the scratches	354
Crackers, starchy	138
Crack of the whip	398
Crazy pendulum	223
water	279
Crooked water surface	190
Crush the jug	262
Crystal palaces	356
grow	332, 351
Cupcakes, brown	324
Curds and chemistry	324
Curie point	216
Cushion, a warm	290
Cut-ups, compass	209

D

Dance, a water	283
Dark heat	283
spots, why the	294
Defying gravity	374
Detector, bell assembly for	48
flying saucer	47
sensitive static	227
vibration	38
Dime, bernoulli with a	95
Dim the lamps	208
Diode (1)	45
Diode(2)	46
Dioxide, sulfer	363
Direction of magnetic fields	226
Disappearing crystals, the case of	344
Distilled wood	364
Dizzy, why get	101
Dollar bill puzzle	8
Don't blink	136
cut the glass	317

Double boiling	326
Do you dare	80
Draft, make a	247
Drill, the Indian	38
Drip	179
Driving in the fog	312
Drum, make your own	31
tuning the	27
Dry solids—contact angles	183
Dyes, natural	116

E

Earth magnetism	217
earth, minerals from the	113
Earthworms	148
Easy battery tester	219
crushed can	265
hammering	12
oxygen	336, 361
Effect, Mr. Doppler's	26
the Tyndall	299
Egg, how fresh	329
Egg & gelatin	327
candling	119
sulfur in your	356
the swollen	334
Elastic air	263
soap bubbles	247
Elastic coins	7
Elasticity	379
Electoscope, simple	211
Electric girl	220
Electricity from a lemon	217
from heat	234
Electrolytes	383
Elevator cord	253
Engine, make a light	60
oil-drop	156
Equilibrium, physical	189
Eradicator	363
Erosion of soil	154
sheet	176
Erratic ball	261
Evaporation & sweating	162
trick	168
Expansion, air	265
Experiment, a jam	335
the great redi	142
William Gilbert	236
Explosions on the kitchen stove	273
Eye games	137
trick	142
trick of	141
Eyeglasses, how to clean	317

F

Fallacy, the balloon	269
Falling, funnel	166
leaves	125
rain or snow	124
Fast cotton	281
Faucet, water stream from a	164
Fermentation, carbon dioxide by	339
Fields, magnetic	229
Finger, freeze to	285
a sausage from the	291
Fingertips, smoke from the	70
Fire, quench the	314
Fireproof, handkerchief	276
paper	287
Fireworks from a lemon	74
Fisherman's puzzle	166
Flame, the ghost	271
water in	171
Flip the penny	75
Flour, burn	337
Flowers, bleach	341
Fluidics	388
Fluorescent light	304
Fog, driving in the	312
Fouvault's pendulum (1)	393
pendulum (2)	394
Fountain, chicken	243
in a jug	268
Freeze to finger	285
Freezing	278
of tissue	112
Friction	386
of water	381
stick-slip	389
Fruit, swelling	187
Funnel falling	166

G

Games, eye	137
Garden, a salt	332
grow a fungus	105
Gelatin & egg	327
Geotropism	139
Ghost light	99
Glamorizing the wedge	376
Glass, don't cut the	317
harmonica	53
the gliding	258
the hanging	246
Glasses that magnify	293
Good to eat, a weed	322
Grass, hot	109
Gravity, center of	9
defying	374
Greenhouse, an indoor	108
Green leaves in winter	123
Grow a fungus garden	105
a pine tree	129
crystals	332
Glowing in the dark	230

H

Habit, muscle	101
Hammer-head	389
Hammering, easy	12
Hand, a hole in the	70
meteors in the	397
Handkerchief, fireproof	276
Handle, loose	13
Hanging glass	246
Hardened wax, remove	328
Harmonica, glass	53
Hazard, a fire	344
Hear through the teeth	20
Heartbeats through a loudspeaker	29
Heat & cold, reflection of	290
dark	283
Hindu magic	74
Holding the breath	126
Holes, boat of	164
in water	150
that hold	194
Homemade compass	220
telephone	41
Horn, an air	26
Hot grass	109
or cold	102
water, how the	288
How far the storm	15
fast are your signals	131
fresh the egg	329
pure is water	160
the hot water	288
to clean eyeglasses	317
Hummer, tricky	15
Hydraulic jump	174
Hydrotropism	139

I

I beam idea	392
Ice, mystery	274
salt	353
Ice cubes, clear	185
mysterious	170
Icy temperature	282
Illusion, an optical	79
Image, the double	297
Induction, magnetism by	211
Inertia & momentum	7
Infection	145
Ink eradicator	363
Inside-out balloon	244
Invisible writing	91
Irregular vibrations	19
It's magic	72

J

Jar killer, make a	147
lid, the stuck	289
sweating	153
Jet-propelled boat	369
Jug, balloon in	240
cloud in a	188
crush the	262
fountain in a	268
Jump, hydraulic	174
Jumping beans	73

K

Kaleidoscope, make your own	43
Kitchen stove, explosions on the	273
Knives, a bridge of	95
Knife & fork chimes	17

L

Lamps, dim the	208
Layering	119
Layers, liquid	181
Lead chemistry	366
Leaves, falling	125
Leaves, new plants from	114
Leaves, two alike	100
Leaves, water from	101
Lemon, electricity from	217
fireworks from a	74
Leverage & muscle	372
Lifting, weight	375
Light box, a trick	60
Light & chemistry	340
Light, fluorescent	304
Light from the teeth	91
ghost	99
polarized	301
sodium	312
strobe-like	303
waves	305
Liquid layers	181
Liquids, shimmering	161
Little finger, a strong	264
Live wire	192
Lively soap	158
Loudspeaker, heartbeats through a	29
Lumpy custard	318
Lung power	241

M

Machine, a wave	55
Magic	
aluminum	231
board	228
bubble	226
Hindu	74
paper	86
pencil	76
propeller	40
wedge	92
Magnet	
a loop	225
a pencil	224
broken	210
in the bathroom	237
make a big	234
Magnetic	
fields	229
fields, direction of	226
fields, permanent pictures	232
or not	204
sugar	190

Magnetism		
by induction	211	
what conducts	222	
Magnetize the scissors	213	
Magnetizer	55	
a smaller	58	
Magnify, glasses that	293	
Make		
a big magnet	234	
a draft	247	
a jar killer	147	
a light engine	60	
a salt	354	
a tester	216	
carbon dioxide	339	
finger paint	321	
ink	347	
sour milk	323	
tracing paper	321	
your own alcohol lamp	41	
your own drum	31	
your own kaleidoscope	43	
Manufacturer, worlds largest	111	
Marble		
the rolling	399	
tricky	395	
Marriage of the water drops	156	
Mass, velocity, momentum	10	
Match		
box drop	89	
moving	71	
the kitchen	368	
the safety	369	
Measure air moisture	42	
Measure of distance	52	
Medium, mold-culture	105	
Megaphone	33	
Melting		
ice, mystery	170	
under pressure	156	
Merging streams	158	
Meteors in the hand	397	
Milk bottle octopus	238	
Mind, a trick of the	104	
Minerals from the earth	113	
Mold		
bread	104	
culture medium	105	
Molecules, muscular	13	
Momentum		
angular	388	
& inertia	7	
mass, velocity	10	
Money power	205	
More		
elasticity	380	
rolling	400	
sand & water mystery	197	
Mortar & concrete	367	
Motor		
a heat	252	
the psychic	49	
Moving		
matches	71	
toothpick	195	
Mr. Doppler's effect	26	
Mud, wet	191	
Multiplied muscle power	375	
Muscular molecules	13	
Muscle		
habit	101	
leverage	372	
power, multiplied	375	
Music, soda-straw	23	
Musical		
balloon	22	
rubber band	20	
Mysterious		
ice cubes	170	
needle	154	
Mystery		
a bird	201	

a magnetic	76	
colors	308	
cloth & sponge	159	
a color	352	
a gypsy	71	
ice	274	
more sand & water	197	
of melting ice	170	
salt shaker	77	
sand & water	197	
skin diver s	291	
sound	21	
spool	391	
the Hislch tube	65	
water & weight	160	
Myth, a magnet	214	
N		
Natural dyes	116	
Needle		
bashful	209	
dip	206	
mysterious	154	
New plants from leaves	114	
Newton s third law	391	
Noisy can	24	
O		
Odor, a big bad	343	
Octopus, milk bottle	238	
Oil-drop engine	156	
Oily smear	153	
On-shore and off-shore winds	59	
Ouija board	79	
Oxidize iron	347	
Oxygen	346	
easy	336, 361	
P		
Pain, a pinch for	113	
Palaces, crystal	356	
Paper		
burning	287	
& coin trick	242	
chromatography	362	
fireproof	287	
magic	86	
make tracing	321	
stubborn	162	
the rising	249	
Parachute	253	
Pencil, magic	76	
Pendulums	35	
an electric	233	
crazy	233	
Foucault's (1)	393	
Foucault's (2)	394	
the compound	386	
Penny, flip the	75	
Permanent pictures		
of magnetic fields	232	
Phototropism	141	
Physical		
change of a chemical	356	
equilibrium	189	
Pickled eggs	319	
Pickles	315	
Picture transfer	88	
Pine tree, grow a	129	
Ping-pong ball, the goofy	203	
Pin-hole vision	305	
Pipe, open & shut	19	
Plants		
and temperature	133	
force	140	
Polarized light	301	
Pollution		
air	259	
to order	343	
Pop-gun, potato	50	
Potato		
pop-gun	50	

Pot holders	316	
Pouring cold air	256	
Power, lung	241	
Precipitation	361	
Pressure of swelling seeds	143	
Print, a leaf	78	
Problem		
space	336	
weighty	200	
Propeller, magic	40	
Protection		
Cathodic (1)	358	
Cathodic (2)	359	
Puddle, a rainbow	307	
Puffballs	145	
Pulleys	392	
Puzzle		
a tricky	82	
dollar bill	8	
fisherman s	166	
surface tension	194	
Q		
Quench the fire	314	
Question, an old freezing	274	
R		
Rain or snow, falling	124	
Rainbows		
in a bubble	312	
in a puddle	307	
in a record	313	
Rays, sun's	308	
Record, rainbows in a	313	
Reflection		
the fat	295	
under water	311	
Reflector		
a rock	298	
the corner	296	
Remove		
hardened wax	328	
warts	128	
Resistance, wind	257	
Rheostat, a better	68	
Rheostat	66	
Rice puffs, acrobatic	232	
Ridge, reynolds the	173	
Rising		
bubble	192	
water	184	
Rolling right along	382	
Root		
hairs, see	143	
watch the	133	
Rubber band, musical	20	
Rubber eraser	369	
Rug power	212	
S		
Sailing	259	
Salt		
make a	354	
salt garden	332	
separate	332	
shaker mystery	77	
& vinegar on copper	334	
Sand		
castles	182	
the absorbing	116	
water mystery	197	
Scale, build your own	43	
Scissors, magnetize the	213	
Scratches, cover the	354	
Secret writing	72	
See		
root hairs	143	
saw, the candle	373	
the pulse beat	115	
Sensation, an unusual touch	143	
Sense of touch	100	
Sensitive static detector	227	

404

Separate salt & sand	332
Shadow, candle flame	313
Shape of poured water	199
Sheet erosion	176
Shimmering liquids	161
Shocker	
a safe	63
a simple	61
a stronger	62
Signals, how fast are your	131
Simple electroscope	211
Siphon without sucking	175
Skin diver s mystery	291
Sleeve,the warm	279
Slinky, waves shown by	37
Smear	
oily	153
soapy	152
Smoke	
from fingertips	70
ring cannon	272
ring in water	190
Snow, blowing	169
Soap, lively	158
Soapbubbles, elastic	247
Soapy smear	152
Soda	
straw atomizer	238
straw music	23
Sodium light	312
Soggy pancake	318
Soil, erosion of	154
Solubility	350
Solutions	349
Solvents	366
Sound	
mystery	21
what can carry	22
Space problem	336
Spallanzani	127
Spider webs, capture	149
Spiraling water	189
Spool, the mystery	391
Spurt, water	177
Squeeze bottle	276
Starchy	
bubbles	280
crackers	138
Stars, water	302
Static electricity and the TV	235
Sterilize soil	131
Stick-slip friction	389
Sticks, stuck	163
Storm, how far	15
Straw, the obedient	215
Streams, merging	158
Strobe-like light	303
Stubborn paper	162
Stuck sticks	163
Sublime, substances that	368
Substances that sublime	368
Suds	179
Sugar, magnetic	190
Suggestion, power of	148
Sulfur	
dioxide	363
in your egg	356
Sun's rays	308
Supersaturation	358
Surface	
tension	191, 193
tension puzzle	194
Sweating	
evaporation	162
jars	153
Swelling	
fruit	187
seeds, pressure of	143
Swing, the	400
Sympathic vibrations	32, 93

T

Tasteless coffee	118
Technicolor cabbage	335
Teeth	
hear through	20
light from	91
Telephone, homemade	41
Telescope	309
Temperature	
and plants	133
icy	282
Tension, surface	191, 193
Test	
a chemical	364
a potato	202
for vitamin C	146
the flame	365
Tester	
easy battery	219
make a	216
Thermometer, a cricket	109
Think warm	121
Those dark streaks	315
Tissue, freezing of	112
Toaster, chemistry in the	364
Tongue foolers	134
Toothpick, moving	195
Touch, sense of	100
Tree force	107
Transfer, pictures	88
Transverse wave motion	377
Trick	
a balloon	77, 255, 240, 254
a Bernoulli	258
a chemical	92
a fork & spoon	97
a gravity	94
a paper	73
a pepper	69, 73
a salt	96
air pressure	265
an eye	135
an impossible	85
another rubber band	285
another straw through potato	90
balloon	240
bottle	242
brick	10
evaporation	168
of the eye	141
paper & coin	242
potato through straw	90
two mirror	296
Tricky	
colors	310
hummer	15
marbles	395
switches	207
Third law, Newton's	391
Tuning the drum	27
Two	
frictions	274
leaves alike	100
mirror tricks	296

U

Umbrella, water	186
Under pressure, melting	156
Under water, reflection	311
Unusual touch sensation	143
Uphill roll	398

V

Valves in the arm	118
Vibrations	
irregular	19
sympathetic	32, 93
through a cone	27
voice	30
Vision, pin-hole	305
Vitamin	
change	329
C, test for	146
Voice vibrations	30
Volcano, a carbon	363
Volume, a trick in	331
Vortex, why the	199

W

Wacky wick	155
Wall, water on the	161
Warts, remove	128
Watch the roots	133
Water	
a candle burning in	82
bullets	167
cavities	181
clinging	178
crazy	279
drops, marriage of	156
from leaves	101
holes in	150
how pure is	160
in flame	171
on the wall	161
race	316
shape of poured	199
smoke ring in	190
spiraling	189
spurt	177
stars	302
stream from a faucet	164
surface, crooked	190
surface, stretch	150
the wayward	204
umbrella	186
weight mystery	160
wetter	152
Waves	
light	305
motion, transverse	377
shown by slinky	37
Wedge	
glamorizing the	376
magic	92
lifting	375
of air	244
weight hot	277
Weighty problems	200
Wet mud	191
Wetter water	152
What	
can carry sound	22
conducts magnetism	222
holds up the clouds	159
Wheel	383
Which	
falls first	395
one's which	87
Whip, crack of the	398
Whistle, the ghost	34
Why	
get dizzy	101
no blow-up	249
the alcohol rub	290
the dark spots	294
the round drops	157
the vortex	199
Wick, wacky	155
William Gilbert experiment	236
Wind	
blows, the way the	255
resistance	257
on-shore & off-shore	59
Winter, green leaves in	123
Wire, live	192
Wood, distilled	364
Worlds largest manufacturer	111
Writing	
invisible	91
secret	72

Y

Yeast and carbon dioxide	146
Yogurt anyone	319
You and a horse	374